D1496276

HARVARD EAST ASIAN MONOGRAPHS

83

GROWTH AND STRUCTURAL CHANGES IN THE KOREAN ECONOMY 1910–1940

GROWTH AND STRUCTURAL CHANGES IN THE KOREAN ECONOMY 1910–1940

by Sang-Chul Suh

Published by
Council on East Asian Studies
Harvard University

Distributed by
Harvard University Press
Cambridge, Massachusetts
and
London, England
1978

This book is produced by the John K. Fairbank Center for East Asian Research at Harvard University, which administers research projects designed to further scholarly understanding of China, Japan, Korea, Vietnam, Inner Asia, and adjacent areas. These studies have been assisted by grants from the Ford Foundation.

Library of Congress Cataloging in Publication Data

Suh, Sang-Chul, 1935–
Growth and structural changes in the Korean economy, 1910–1940.

(Harvard East Asian monograph; 83)
Bibliography: p.
Includes index.
1. Korea–Economic conditions. 2. Korea–History–Chosen, 1910–1945. I. Title. II. Series.
HC467.S83 330.9'519'03 77-13885
ISBN O-674-36439-2

To My Mother

FOREWORD

A number of books have been published recently describing and analyzing South Korea's economic "miracle," the rapid growth that has quadrupled Korea's GNP over the last decade and a half. Most of these works begin with the early 1960s; a few go back to the 1950s or late 1940s; but none deal systematically with the 35-year period of Japanese colonial rule. And yet a great deal happened to Korea's economy during that colonial period, and it is doubtful that one can really understand the current economic achievements without some knowledge of what went before.

As Professor Suh demonstrates, a great deal did happen to Korea's economy under the Japanese. Most colonial regimes introduced orderly and efficient administration, but the Japanese went much further. They brought in modern techniques in a generally successful effort to raise rice output and exports. In later years the Japanese also began to introduce modern industry, including producer goods industry. Per capita commodity product (the closest one can get to per capita GNP) rose substantially.

Korean economic growth under the Japanese was a great success story except for one disturbing fact. Very few Koreans benefited materially from that growth, and the great majority actually experienced a decline in their standard of living. One frequently hears charges that growth benefits the rich at the expense of the poor, but most such charges fall into the category of unsupported political rhetoric. Professor Suh, in contrast, documents the charge with analysis supported by data. The main beneficiaries of Korean growth at that time were Japanese consumers who ate inexpensive Korean rice and the Japanese military machine that used Korea's railroads and some of its industries (plus considerable amounts of manpower).

If Korea's growth from 1910 to 1940 was mostly by and for the Japanese, that growth still left a legacy, some of which played a positive role in economic events once the Japanese had gone. An understanding of the nature of that legacy is one of the several contributions of this book.

Professor Suh received his B.A. and M.A. from Clark University and Ph.D. from Harvard University. After teaching for a number of years at Clark University and the Economic Development Institute of the World Bank, he returned to Korea in 1972, where he is currently a Professor of Economics at Korea University.

Dwight H. Perkins
John K. Fairbank Center for
East Asian Research

CONTENTS

PREFACE

This book attempts, first of all, to measure the rates of economic growth and structural change in the period 1910–1940. Economic growth is measured by the estimates of commodity-product along with the expansion of the labor force and foreign trade. The term "commodity-product" refers to net output originating in agriculture, forestry, fishery, mining, and manufacturing. The scarcity of data is the main reason for using the commodity-product concept instead of national income concept. Structural changes are measured by the industrial distribution of commodity-product and labor force as well as by the changes in organization and regional pattern of producing units in each industrial sector.

This book also attempts to evaluate the quantitative findings in comparative perspective to ascertain the unique features of the growth pattern during the colonial period. The systematic organization of quantitative data consistent with modern economic concepts permits cross comparison. The quantitative findings are often supplemented by qualitative materials so as to give historical perspective to the whole study. In this way, it is hoped to show the relationship of colonial development to the postwar economic growth of Korea.

While most scholars agree that fundamental changes took place in the Korean economy during the colonial period of 1910–1945, the implications of the economic change for Korea remain highly controversial. I have attempted to shed some light on these controversial issues by producing a work in quantitative economic history. The war period of 1941–1945 is not covered here because the normal pattern of colonial development was disrupted by the impact of World War II.

My work on the study of economic growth during the colonial period began early in the 1960s for a doctoral dissertation at Harvard University. This book is a revised version of that doctoral dissertation. The revision was mostly done while I was a faculty member of Clark University during 1964–1969. I wish to acknowledge my debt to the National Science Foundation for a faculty research grant in 1967–1968, during which time I was on leave from Clark University to work on the extensive revision. My trip to Japan and Korea during that period to collect additional materials was partly financed by a grant from the Social Science Research Council and the American Council of Learned Societies.

I am deeply indebted to Professor Simon Kuznets, under whose guidance my doctoral dissertation was written. He also read one draft of this monograph. Without his generous advice and criticism, this book would not have come to fruition. I am also indebted to Professors Henry Rosovsky and Dwight H. Perkins who read many drafts both of the doctoral dissertation and this book in its prepublication stages.

I also wish to acknowledge my debt to the Korea Development Institute which provided me the opportunity to complete the final draft of this monograph as a visiting fallow upon my return to Korea in 1972. Professor Leroy P. Jones of Boston University kindly read the entire manuscript while he was a visiting fellow at Korea Development Institute and offered invaluable comments which have improved the presentation.

I am aware of possible errors in the estimates of commodity-products and the use of quantitative data, for which, needless to say, I assume full responsibility.

Sang-Chul Suh

July 1977
Seoul, Korea

Chapter I

THE HISTORICAL BACKGROUND

It is often said that the modern period in Korean economic history began in 1876 when a commercial treaty was signed with Japan, followed by similar treaties with other nations in later years. These events marked the termination of a centuries-old policy of seclusion. Thenceforth, international contacts were made on a continuing basis, particularly with Japan. However, the mere opening of Korea's doors to foreign countries failed to generate the pressures necessary for the economy to leap into modern economic growth. Instead, the Japanese influence penetrated deeply into the country, and Korea was eventually annexed to Japan in 1910. In view of these events, the transitional[1] period in this study may be selected as 1876–1910.[2]

Korea's Transition to the Colonial Period, 1876–1910

What facets of the economy during the transitional period should be examined? We may concentrate on the factors that we know to be important for our purpose during the colonial period. In particular, we shall first examine briefly the process by which the Yi dynasty was replaced by the colonial administration of Japan (The Government-General of Chosun), followed by a survey of major economic changes during the transitional period.

The Road to Colonialism[3]

In the late sixteenth and early seventeenth century, Korea experienced devastating wars with Japan and with the Manchus in China. This period of agonizing wars was followed by a long period of peace, during which a seclusion policy was pursued until 1876.

In retrospect, there is no doubt that this isolation engendered many obstacles to modern economic growth in Korea.[4] First, this policy prevented international contacts which, in the West, often played a vital role in modern economic growth during the nineteenth century. What is more, the lack of foreign contacts made it even more difficult for the Korean monarchy to recognize an increasing economic backwardness compared to the West and Japan during this period.[5] Second, the isolation policy (and the long period of peace) meant a loss of the traditional source of strength for political unity of the nation. Being surrounded by powerful nations (China, Japan, and Russia), Korea traditionally had to bear a heavy burden of national defense. This required the unified military force of all regions within the country; any regional force alone was simply not strong enough to prevent foreign aggression. This practical necessity of national unity for defense gradually diminished during the period of peace.

When the political unity of the nation was dissipated, the effectiveness of the central government was substantially undermined, and social disorders developed. The land-tenure system[6] of the Yi dynasty was highly vulnerable to the inefficiency of the central government. The tenure system might be called "quasi-feudal" in that all the arable land was in principle in the possession of the monarch. Under the system, the arable land was cultivated by peasants for their landlords (*yangban*), who in turn payed an "in-kind" land tax (which included rent elements as well) to local administrators representing the central government. When the efficiency and power of the monarchy declined, many landlords took advantage of the circumstances by imposing heavy burdens on their tenants, and there were numerous incidents of civil servants' abuses and appropriations of taxes.

While domestic conditions were deteriorating, there were some basic changes in the international affairs of Asia during the late nineteenth century, most notably the penetration of Western political influences such as the prevalence of colonization. Following the Meiji Restoration in 1868, Japan also started to mobilize her power toward the aquisition of colonial territory in Asia. Under these circumstances, a certain degree of military power was required to maintain a national policy of seclusion. Korea failed to meet this requirement because of a long period of dynastic friction, social disorder, and economic stagnation.

It is not surprising, therefore, that the abrogation of Korea's seclusion policy did not emerge from within but was imposed upon Korea by the military force of Japan. The major events that led to the signing of the first commercial treaty with Japan in 1876 can be summarized in the following way:

> After 1860 many powers showed a great interest in the Hermit Kingdom (Korea), and a number of incidents arose because the Koreans were determined to keep out all foreigners. In 1875 a Japanese gunboat was engaged in surveying the mouth of the Korean river *Han*, which it had no right to do. The ship was first fired upon by Koreans and this this was regarded as an incident justifying the dispatch of Japanese gunboats and military transports into Korean waters to enforce certain demands. The result was a Korean-Japanese treaty of commerce, signed on February 26, 1876.[7]

In this way, the country was forced to carry out an open-door policy while lacking both a strong government and a solid economic foundation. For this reason, the mere exposure of the country to foreign countries created even more dynastic friction and disunity, now supported by foreign powers (particularly Japan, China, and Russia), along with the rise of foreigners' economic activities.

It was also during the period of the open-door policy that a new intellectual movement, *Donghark*[8], gained a strong appeal among the discontented elements of society. The underlying principle of the *Donghark* movement was to achieve human equality through reforms. This movement was a significant departure from the traditional ideology of Korea in respect to advocating reform of Korean society and seeking to reverse the tide of the increasing penetration of foreign nations into domestic affairs. In retrospect, it may be said that this movement had the potential to enlist all discontented and patriotic groups in a nationwide revolt against the declining government in 1894. When the *Donghark* revolt against the Yi dynasty took place, however, the Japanese military force quickly came to the government's aid in suppressing it. The reform movement emerging from within was halted by the military force of Japan and brought Korea a step closer to eventual colonization by Japan.

The political influences of China and Russia were eventually eliminated from Korea through Japan's victories in the Sino-Japanese War (1894) and the Russo-Japanese War (1904–1905). Korea became a Japanese protectorate in 1905, when Korea agreed to follow Japan's advice on matters of reform, public finance, and foreign affairs. The official annexation of Korea by Japan took place in 1910.

Major Economic Changes

The immediate effects of the commercial treaty with Japan in 1876 were a rapid expansion of foreign trade and its complete monopoly by Japan. Within six years following the treaty, the total value of trade between the two countries showed an increase of more than twenty times over that of 1872–1875.[9] While Japan's monopolistic position in Korea's foreign trade was challenged by other nations during 1881–1904,[10] Japan's victory in the Russo-Japanese War firmly established her monopolistic position.

The major reasons for Japan's interest in trade with Korea may be seen in the industrial composition of commodities traded during the transitional period. Korean exports were largely agricultural products, while imports from Japan consisted mainly of manufactured products (shirts, linen, cloth, manufactured silk, and so forth).[11] Thus, the pattern of trade between the two countries was well suited to the needs of Japan during its industrialization period. It was also consistent with the relative factor endowments of the two economies.

The increasing volume of trade between the two countries had far-reaching effects on the Korean domestic economy during the transition period. The expansion of trade with Japan was the determining factor in the emergence of modern sectors within the Korean economy. In addition, foreign trade set the stage for direct participation of Japanese in the Korean economy, that is, through Japanese ownership of land. Exports of agricultural products from Korea aroused Japanese interest in direct ownership of cultivated areas in Korea. However, the ownership of land by foreigners was not legally recognized until shortly after the transitional period.[12]

No large-scale Japanese immigration took place during the transitional period. The total number of Japanese possessing arable land in Korea was only 3,400 in 1909, and the extent of their holdings was 60,000 *chungbo*,[13] or about 2 percent of the total cultivated area.[14] Despite the legal recognition of foreign ownership of land in Korea, it was well known that Japanese farmers were still reluctant to move in. There were several reasons for this.[15] First, the lives and fortunes of Japanese in Korea were insecure during this time, largely because of the antagonistic attitudes of Koreans toward them. Second, the land system was too complicated, and accurate reports on the location of land were not available. Third, there was a lack of organizations to encourage Japanese immigrants. It was only after the official annexation of Korea that all of these obstacles were eliminated in order to ease Japanese immigration to Korea.

The open-door policy during the transitional period also gave rise to the emergence and rapid growth of some modern sectors in the Korean economy. Let us consider first the establishment of modern financial institutions and a sound currency system. The origin of modern banks in Korea goes back to 1878 when the Dai Ichi Bank of Japan set up a Korean branch in order to facilitate the expansion of foreign trade.[16] This was followed by the establishment of new Japanese and Korean banks. By 1905, there were two Korean banks and three Japanese banks with branches throughout the country.

Modern financial institutions and the expansion of foreign trade gradually increased the use of money throughout the country, replacing the traditional method of barter exchange. The use of money as a means of exchange was further accelerated after the Sino-Japanese War by the substitution of cash tax payments for payments in kind.[17] However, the currency system in Korea was extremely confused during most of the transitional period because there were many types of money with uncertain values. In order to remedy the situation, a Japanese financial advisor was appointed to the Korean government. His primary function was to withdraw from circulation the old Korean coins and to introduce new coins consistent with the Japanese currency system.[18]

The Korean branches of the Dai Ichi Bank were converted to the Central Bank of Korea in 1905, thereby linking the Korean currency system with that of Japan. In this way, modern financial institutions and a viable currency system were introduced for the first time.

It was also during the transitional period that the first heavy investments in social-overhead capital were made, beginning with the construction of railways by Japanese capital and technicians. The major railways constructed during this period were[19] the Seoul-Inchun line in 1900, the Seoul-Pusan line in 1905, and the Seoul-Euiju line in 1906, connecting the south and north parts of the Korean peninsula.

Also, during this period Japanese immigrants to Korea introduced new factories equipped with modern facilities. After the Russo-Japanese War, these developments were particularly noticeable in the manufacturing of tobacco, liquor, and electricity and in the construction of rice mills.[20]

The Colonial Period: Economic Policy

Whatever Japan's basic purpose in annexing Korea might have been,[21] it was a necessary step to achieving complete control of the Korean economy, making it a part of the periphery of Japan. The official annexation of 1910 brought the establishment of the colonial Government-General based on a political system of strict totalitarianism.[22] Under such a political setup, the destiny of the Korean economy was largely determined by the economic policies of the Government-General. Since there was no representation of Korean interests in national policy formulation, the economic policy of the colonial period was always in line with the changing needs of Japan. Therefore, the policy evaluation must be carried out with reference to the changing conditions of the Japanese economy during the period under review. Only in this way will the major factors that shaped the pattern of Korean economy be made clear.

From the standpoint of major policy objectives, it is convenient to divide the prewar colonial period into the following three phases: 1910–1919, 1920–1930, and 1931–1940.[23]

The First Phase, 1910–1919

At the time of Korea's annexation in 1910, Japan had already experienced several decades of modern economic growth. According to Professors Ohkawa and Rosovsky, the initial phase of Japan's modern economic growth took place during 1886–1905:

> The end of the initial phase of MEG (modern economic growth) has been defined as the point at which the growth rate of the traditional economy begins to decline, i.e. 1905. This is the end of a phase because from then on the modern economy could no longer count on the same degree of support from the traditional sectors. While the economic relations between the two sectors persisted, the modern economy had to look more to its inner strength or its export capabilities if rapid growth was to be maintained.[24]

By the first decade of the twentieth century, the Japanese economy was entering an era of industrial capitalism,[25] and foreign markets played an important role in this industrialization. Even before the twentieth century, modern textile exports were firmly established through Japan's monopoly positions in Korea and China resulting from Japan's victory in the Sino-Japanese War.

The agricultural position of Japan had also undergone a basic change by this time. Beginning with the last decade of the nineteenth century, Japan's position shifted from that of having a net surplus of food grains to that of a net importer.[26] However, imports were not a permanent solution to the shortage of food grains because she was faced with balance of payment deficits, which severely curtailed her ability to import consumer goods.

In view of the above needs of the Japanese economy, it is not surprising that the overriding policy objectives of the Government-General during the first phase of the colonial period were to develop the Korean economy as a major source of food supply and to provide a market for Japanese manufactured products. These policy objectives were consistent with the prevailing factor endowments of Korea during that time. Korea was predominantly an agrarian economy, while relying heavily on imports of manufactured products since the open-door policy of 1876. Immediate potential for growth in this sector existed in the form of reclaimable land and unemployed workers.[27]

However, the first phase of the colonial period turned out primarily to represent a preparation for effective control of the economy rather than actual introduction of any significant change. As a prerequisite for colonial development, a new institutional setting consistent with the policy objectives had to be established, and Korean nationalism surpressed. These were indeed formidable tasks.

> When Japan swallowed the ancient Korean Kingdom in 1910, she confronted a situation in some ways quite opposite to that of Taiwan fifteen years earlier. In economic matters Korea's long used soils, deforested hills, and north-temperate–zone climate held no promise of agricultural growth like that of lush semi-tropical Taiwan. Politically, on the other hand, the long tradition of Korean independence and the recent struggle to preserve it nurtured an ineradicable anti-Japanese nationalism. In Taiwan, a rather new Chinese province had been brought under Japanese control before the rise of the Chinese nationalist movement. In Korea, nationalism was already very active among the upper class.[28]

It is outside the realm of this study to examine the various coercive measures taken to implement colonial policies and to establish an effective network of colonial administration, replacing the centuries-old Korean government. It may suffice here merely to point out that Japan attempted to eradicate by military force the national identity as well as the unique culture of Korea. Thus the first phase of the colonial period involved what was often called "the politics of military force."

In this political climate, the major economic measures taken during this phase may now be outlined. First, the Government-General adopted the Japanese civil law in 1912 and undertook an extensive land survey during 1910–1918. The adoption of Japan's civil law provided a legal ground for the private ownership of land, including Japanese ownership in Korea. It also abolished the traditional class relationships based on birth. Henceforth, the landlord-tenant relationship was to be specified by annual contracts. The land survey was carried out to obtain accurate data on Korean agriculture so as to implement the system of private ownership of land and to allow planning for the development of the agricultural sector.

The above measures were similar to those taken in Japan after the Meiji Restoration in order to modernize the traditional structure of the society and to establish the landlord class. However, the measures taken in Korea had one additional purpose, namely, the colonization of the Korean economy. In the absence of employment opportunities elsewhere, the immediate effect of adopting the new civil law was to place tenant farmers under the tighter control of landlords.[29] The land survey was instrumental in enlarging both the landlords' share of total arable land as well as that of Japanese residents in Korea. The new law and the land survey established a rural institutional setting wherein the subsequent development of this sector was achieved by the landlords' drive for expanded production rather than by incentives of the farmers. Japanese landlords were placed in strategic positions in the rural economy to ensure its control. That the new land-tenure system continued to disregard the importance of farmers' incentives as a source of development suggests the preservation of the traditional landlord-tenant relationship. Reforms and improvements by the Government-General were limited in their application to the conditions governing landlords' incentives for the development of the agricultural sector.[30]

The second major economic measure was the enactment of the so-called "Corporation Law" (1911), which insured a monopoly position for Japan's manufactured goods in Korea. The law empowered the colonial government to control (and to dissolve if necessary) both new and established business enterprises. It severely restricted investments in non-agricultural sectors, and discouraged private

Japanese capital inflow into these sectors in order to prevent the establishment of new industries that might compete with industries in Japan. Above all, the Corporation Law was in practice often used to discriminate against Korean business enterprises. When the Korean government and its military force were dissolved at the time of annexation, thousands of the Korean elite had to be relocated. However, their activities in non-agricultural sectors were restricted by the Corporation Law. In this way, the dominant position of Japanese residents in Korea was also established in the non-agricultural sectors of the Korean economy. The percentage share of Japanese in the total paid-in capital of business enterprises increased from 31.8 percent in 1911 to 79.6 percent in 1917 and to 90.3 percent in 1921.[31]

Third, by the completion of currency reform in 1911, the goundwork was laid for economic integration of the two countries. Thenceforth, Japan's currency was acceptable in Korea and the Korea currency in Japan. Complete economic integration, including the abolition of tariffs, however, had to be postponed because the colonial government did not wish to risk losing the confidence of other nations in trade with Korea by an abrupt change in the inherited tariff system. Thus, it was declared that the existing tariff system would be maintained until 1920.

Fourth, the main emphasis in the allocation of public funds was placed on the provision of the infrastructure required for the effective administration of Korea. While the railway trunkline connecting the south and north of the Korean peninsula was completed during the transitional period, various branch lines were added during the first phase. In addition, highway construction, the establishment of telephone and telegraph systems, and harbor improvement were undertaken.

Finally, various measures designed to facilitate Japanese immigration were adopted in 1910, and the Oriental Development Company[32] was assigned the task of organizing large-scale immigratior of Japanese to Korea.

The Second Phase, 1920–1930

It is well known that Japan experienced unprecedented prosperity during the period of World War I. The rapid industrialization and accumulation of capital led to a basic change in Japan's factor endowments: its position was shifted from an economy with a labor surplus to one having abundant capital.[33] The prosperity of the war period, however, also had some negative effects on the Japanese economy:

> Japan experienced inflation in part as a result of the export boom and in part due to the lack of an adequate government financial policy. The results included a badly distorted wage structure with many classes of people experiencing a decline in real income. Food prices rose enormously, and the Rice Revolt of 1918 was an ominous warning to the government.[34]

These changes in the Japanese economy by the end of the first phase of the colonial period shaped the new policy objectives for the second phase in Korea.

Domestic conditions in the late 1910s in Korea also called for some modifications of the initial approaches to the colonial administration of Korea. While the Government-General was quite successful in reshaping the institutional setting, governing by military force tended to incite more anti-Japanese nationalism, rather than to suppress it. This trend was clearly reflected in a series of independence movements culminating in the nationwide movement of 1919, "Samil Woondong." The nationalist movements in Korea led Japan to realize that coercive measures backed by military force could no longer be effective, and a new political development within Japan also advocated a more liberal approach to colonial administration.[35]

Thus the second phase of the colonial period began with administrative reforms designed to replace "governing by military force" with "cultural policy." This marked the beginning of a more positive approach toward achieving assimilation (and eradication of the national identity of Koreans) by cultural means (such as education, training, and so forth) instead of by coercive measures.

First, the Corporation Law was abolished in 1920, thereby

lifting the major restriction on capital flow from Japan. Tariffs on commodity trade between the two countries were also eliminated. Second, an extremely ambitious plan designed to increase rice production was launched. Complete economic integration (thus allowing for regional specialization) and the newly established land-tenure system with the dominance of Japanese landlords were two factors favorable to the undertaking of this plan.

The Third Phase, 1931–1940

This phase of the colonial period witnessed the establishment of large-scale industries in Korea. The major emphasis on rice production during the previous phase was abruptly replaced by concentration on rapid industrialization. Once again, this new development was brought about by the shift of Japan's policy toward Korea in favor of industrial development, originating from the basic changes within the Japanese economy.

The depression of the 1930s brought a new era to the government-business relationship throughout the industrial nations of the world, and Japan was no exception to the trend. To this should be added Japan's special problem arising from the intensification of dualism dating from the beginning of the twentieth century.

> Given the differential structure, it was difficult to strengthen the domestic market on a continuing basis. In terms of gainfully occupied workers, the largest segment of the economy was condemned to lower productivity and low incomes.[36]

On the other hand, the export boom of Japan following its currency depreciation of 1931 was immediately met by a boycott of Japanese products (by other nations) through higher tariffs and quotas.[37] Because of these difficulties, the intervention of the Japanese government in aiding private business was progressively increased, particularly after the radical groups in the army, and their supporters, gained absolute power in national politics.[38] The Japanese government attempted to diversify its industries to create a self-sufficient economy within the Japanese empire. Coupled with the urgent needs of military expansion following the Manchurian Incident of 1931–1932, Japan's economic policy was framed to stimulate the utmost expansion of heavy industries. Thus, this phase of the colonial period coincides with Japan's era of "semi-war economy."

When the necessity arose in Japan to expand the industrial complex beyond its national boundaries, the strategic importance of Korea was recognized for several reasons. In the first place, Korea provided certain cost advantages for establishing heavy industries, including a variety of mineral resources, an abundant supply of hydroelectric power, a cheap labor supply, and so forth. With respect to the labor supply, it should be noted that the success of the colonial policy in Korea made the Japanese language the national tongue in place of Korean. Furthermore, by this time the security of Japanese residents in Korea was fully guaranteed. Second, Japan's acquisition of Manchuria placed Korea in a strategic location wherein it became an important entrepôt for the trade between the two regions while expanding Japanese territory toward China.[39]

Domestic conditions in Korea also required some diversification of industries, a change from the initial policy of monocultural development. Beginning around 1927, the plan for the agricultural development of Korea encountered serious difficulties arising from the agricultural depression in Japan. In order to avoid further deterioration of the rural Japanese economy, it was decided to discontinue the expansion of rice production in Korea.

Thus, the interests of Japan's large business groups, supported by its government and by the colonial government in Korea, found a fertile ground for expanding Japan's industrial complex to incorporate Korea. The colonial government took various measures to induce large Japanese business groups to extend their industrial activities to Korea. Among these measures were a deliberate maintenance of a liberal policy toward business at a time when government control was formidable in Japan; direct aids to big business in the form of subsidies; and the provision of land for the location of new industries. The rate of industrialization was further accelerated when Japan entered into the period of active preparation for World War II.

Chapter II

ESTIMATES OF COMMODITY PRODUCTION, 1910–1940

This chapter attempts to estimate the Korean net commodity value added in constant prices. It would, of course, be preferable to derive quantitative series corresponding to the national account statistics in order to measure the over-all growth and structural changes in the Korean economy. To the best of our knowledge, however, no attempts were made in the past to estimate the national income of Korea for the colonial period, owing to a lack of relevant data.

The primary data available in the official sources cover only the commodity-producing sectors, namely agriculture, forestry, fishery, mining, and manufacturing. Thus the scope of our estimates has to be limited to these sectors. Compared to the concept of national income, our estimates suffer from the omission of the following sectors of considerable importance: construction, trade, services, and public utilities. However, it is an underlying assumption of this study that the limited scope of our coverage will not distort the long-term trend of the Korean economy on the grounds that there is a close correlation between the trends in commodity-product and in the rest of the economy, particularly in low-income countries.[1]

The primary data contain information on the volume and market value of commodity-production. While we cannot make any extensive cross-check on the accuracy of the primary data at this time, it may be pointed out that the colonial administration in Korea had a relatively good reputation for compiling statistical data on commodity-production.[2]

The market value of commodity-production in the official data includes the cost of all intermediate goods, which is necessarily higher than the net value added. This is why we need to adjust market value to derive estimates of net value added in commodity-production. The ideal procedure for the derivation of net value added would be to deduct the cost of intermediate goods from market value. However, the lack of cost data renders such a computation impossible. Some indirect methods have to be devised to derive gross value added. The details on the methods and data used for the derivation will be presented in the following sections.

By way of introduction, it may be useful to outline the general approach. First, the official data are classified by industrial origin and adjusted for consistency and reliability to yield market value (MV). Second, the standard ratios of net value added to market values are estimated for each industrial sector (R), and these ratios (called net product ratios) are applied to the market values of commodity-products to derive net value added (NVA)

$(NVA)_{ij} = R_{ij} \times (MV)_{ij}$, where i is the industrial sector, and j is the specific year.

Agricultural Production

Market Value

Until an extensive land survey was completed in 1918, even the colonial government did not claim much accuracy for the previously published data on agricultural production. Upon completing the land survey, the government revised the agricultural statistics of 1910–1917, using the 1918 data as benchmarks.[3]

It seems reasonable to assume that the data obtained from the land survey are fairly accurate. The completion of the survey took eight years (1911–1918) and cost over K¥ 24 million (equivalent to almost one year's tax revenue, 1917–1918). The survey was undertaken in response to the urgent need of the colonial government for accurate data about agriculture during the early years of the colonial period, and there were practically no limits on the use of coercive measures to implement the objectives of the colonial government during this period. At the same time, there were no strong incentives for concealment or underreporting of cultivated areas, because a new system of private land ownership was established through the land survey.[4]

Following the completion of the land survey in 1918, the official data of agricultural production for the subsequent years were compiled, based on the reports coming from the lowest level of colonial administration (village clerks).[5] These reports represented largely the judgments of village clerks concerning crop conditions in their villages each year without any actual survey of crop conditions. These rather superficial judgments probably resulted in errors in the official data.[6]

The method of compiling the official data on rice production was drastically changed in 1936. The new method required an actual survey of rice production, replacing the "desk work" of village clerks in the previous years.[7] This change must have improved the reliability of official data from 1936. We have two figures for the 1936 rice production: 19.4 million *suk* by the new method and 15.4 million by the old method. The discrepancy of 4 million *suk* may be taken to represent the magnitude of underreporting caused by the old method. This suggests that the official data on agricultural production must be modified to account for the possible underreporting of rice production prior to 1936.

In estimating the magnitude of underreporting, we assume first that the land survey data in 1918 and the 1936 survey data provide reliable information on rice production. Using these bench-mark surveys, our task is to find a reasonable basis of interpolation for the intervening years.

In order to examine the possible sources of underreporting, the process of data collection must be scrutinized. It begins with the reports of village clerks concerning rice production. The judgments of village clerks on crop conditions, in the absence of any actual surveys, must have been influenced mainly by information gathered from landlords and farmers, and by referring to the allegedly accurate data of the 1918 survey. The 1918 data, however, became increasingly obsolete as many years of rapid agricultural development ensued. Accordingly, the official data on rice production for the later years must have been based increasingly on the information of crop conditions provided by landlords and tenant farmers alone.

Thus, it is important to examine whether there were any incentives for the underreporting of rice production either by landlords or by tenant farmers. As for the landlords, we have some reason to doubt their motives for underreporting. Above all, the burden of land taxes was determined according to the fixed value of land (called its "legal value") set at the time of the land survey. And there was no plan for any land revaluation during the later years of the colonial period. The colonial policy of Japan in Korea was to make the ownership of land as profitable as possible. Thus, government levies were not considered to be a heavy burden on landlords.

Unlike the landlords, however, evidence seems to suggest the existence of strong incentives for underreporting by tenant farmers. The amount of rent paid by them was commonly proportional to the amount harvested, and this encouraged underreporting. As Chapter V will explain, rice production was the major source of family income for tenant farmers. Hence, some degree of underreporting by them was often necessary for the very survival of farm-households at times of poor harvest and extremely high rents. Since the land-tenure system introduced by the colonial government did not guarantee any minimum level of food supply for tenant farm-households, the "cheating" of tenant farmers was often considered to be "a necessary evil of life." In the case of the Korean tenant farmers under Japanese landlords, underreporting of rice production might have been encouraged by nationalistic feelings.

On the basis of these observations, we assume that the underreporting of rice production in the official data was largely attributable to the tenant farmers and that the magnitude of underreporting was positively correlated with the degree of hardship imposed on them. In the absence of relevant data, the degree of hardship may be measured by per capita availability of rice, because rice production was the major source of income for farm-households.

Under the above assumption, the following method is used to adjust for the underreporting of rice production in the official data. We first select a level of per capita rice consumption that may be regarded as the lower limit prevailing during the colonial period. Then, the magnitude of underreporting may be approximated by the discrepancy between this lower limit and the level of per capita rice consumption derived from the official data for 1919–1935. In selecting the lower limit, it seems appropriate to use the lowest level of per capita rice availability during the active war-years of 1941–1945, which was roughly 0.55 *suk* in 1944. According to this calculation, underreporting is involved only for 1924–1935 as shown in Table 1.

The plausibility of our estimates in Table 1 may now be tested. The estimate of 1936 rice production by our method of adjustment approximates very closely the official data obtained from the new survey method. Another test of plausibility may be made by comparing the growth trends between the pre-survey periods of 1920–1935 and 1936–1941. Since there were no basic changes in the agricultural development policy during the two periods, one would expect a fairly smooth growth curve connecting them. For the purpose of comparison, we present the following three growth curves (log-linear form):

1. 1936–1941	Official Data (by New Survey)	$Q=4.32932+0.01209T$	$R^2=0.998$
2. 1920–1935	Official Data	$Q=4.14179+0.00611T$	$R^2=0.997$
3. 1920–1935	Adjusted Data (Table 1)	$Q=4.12517+0.01289T$	$R^2=0.994$

The comparison of (1) and (3) gives a fairly smooth growth trend between the two periods, confirming our expectation, whereas the growth trend between (1) and (2) shows a kink which probably reflects the effects of underreporting during 1920–1935. The adjustment seems to be satisfactory in eliminating the effects of underreporting.

As for the market values of other agricultural products, there is no evidence of underreporting (or overreporting) involved in the official data. The hypothesis we developed in connection with rice production should not be applied here, because other agricultural products were not the major source of income for farm-households. Thus we propose to use the official market values without any

Table 1

VOLUME OF RICE PRODUCTION
(in million *suk*)

Year	Official Data (A)[a]	Revised Estimates (B)	(B)-(A)/(A)
1924	13.2	14.2	7.1%
1925	14.8	15.3	3.3
1926	15.3	16.1	5.0
1927	17.3	17.9	3.5
1928	13.5	16.0	8.7
1929	13.7	16.1	17.5
1930	19.2	20.1	4.7
1931	15.9	19.1	20.4
1932	16.3	19.7	20.3
1933	18.2	21.5	18.3
1934	16.7	20.8	24.4
1935	17.9	21.7	21.5
1936	15.4[b]	19.3	25.4

a. The figure obtained by the old method is used to compare with the one from the new survey method (19.3 million *suk* in 1936).

b. Taken from *Chōsen tōkei nempō*, 1924–1936.

adjustments. Our revised estimates of rice production alone, however, will significantly improve the reliability of agricultural statistics as a whole, owing to the dominant position of rice in the total agricultural product. Table A-1 presents the market values of all commodity-products originating in agriculture and forestry.

Net Value Added

The method used to derive net value added from the market value of agricultural output may be summarized in the following way.

We first estimate the net product ratio for a base year. The base-year ratio is then modified by the annual variation of the indicators selected, yielding the net product ratios for the entire period. Finally, net value added is estimated as the market value multiplied by the net product ratio. Since the information on intermediate goods is available only in the case of rice production, the net product ratios of all the crops are estimated using the net product ratios of the production. This procedure may be justified in view of the predominant position of rice production in total crops (over 50 percent during the entire period).

We selected 1933 as the base year because an extensive survey on the cost of rice production was taken in 1933. The survey data are shown in Table 2. For the derivation of the value-added ratio, the cost of rice production must cover such intermediate goods as seeds, fertilizers, materials, and tools.[8] The total cost of intermediate goods per *suk* of brown rice in 1933 amounted to K¥4.79, while the average market price was K¥20.41.[9] Thus the value-added ratio for 1933 was .765.

Our next task is to estimate the annual variations in the value-added ratios. In order to maintain consistency in our estimates, the following assumptions are made. First, that the unit cost of rice production in current prices is a function of fertilizer price changes and the quantity of fertilizer used per unit area.[10] Second, that the annual variations in the market value of rice production are determined by the changes in rice price and average yield per cultivated area. Third, that the upper limit of the value-added ratio (when no fertilizer is used) is assumed to be .90.[11]

Under the above assumptions, the annual variations of value-added ratios from the base-year ratio may be computed as follows:

$$\underset{(\text{in } \%)}{VAR_i} = 90 - 13.5 \frac{a_i \times b_i}{A_i \times B_i}$$

where VAR_i is value-added ratio of the ith year, a_i is the index of fertilizer use per unit area,[12] b_i the price index of fertilizer, A_i the index of rice yield per unit area, and B_i the price index of rice. Table A-2 presents our estimates of value-added ratios along with the various indices used for the calculation.

As for other agricultural products including forestry, we did not have the empirical data required to estimate the value-added ratios. In this connection, it may be noted that the agricultural

Table 2

COST OF RICE PRODUCTION, 1933
(per one *suk* of brown rice)

Items	Cost in K¥
Seeds	.33
Wages	7.15
Fertilizer	3.25
Materials	.77
Buildings	1.01
Tools	.44
Taxes & other levies	1.65
Interest or payment to landlords	7.13
Processing of rice	.50

Source: Chōji Hishimoto, *Chōsen-mai no kenkyū*, (Tokyo, 1938), p.235.

policy of the colonial period placed major emphasis on the expansion of crop production, with the non-crop production lagging behind. The value-added ratios of non-crop products must therefore have been higher than those of crop production, reflecting the lion's share of the payments to the agricultural sector itself in the total cost of non-crop production. We thus used the upper limit in the value-added ratios of crop production (.90) as a uniform ratio of value added to market value for all non-crop products.

Even if full allowance is made for the crudeness of our estimates of value added for non-crop products, the use of the separate value-added ratios between crops and non-crop products will nevertheless improve the over-all estimates of agricultural production by showing the effects of structural changes in agricultural output. Our estimates of net value added originating in agriculture and forestry are presented in Table A-3.

Manufactured Production

Market Value of Output

There are two sets of series on the current market value of manufactured output in the official data. One covers "factory output" and the other shows the total output of what may be called "manufactured commodities." The official definition of a "factory" was a production unit employing at least five workers with power (electric or steam) facilities.[13] The data on "manufactured commodities" include all the household-industry output and factory output except rice processing, cotton processing, lumbering, and metal (other than iron) refining. Let us first estimate the market values of factory output, followed by household-industry output.

In order to make adjustments to the official data on factory output for the omission of the government-owned factories, we attempted to derive the relevant data from the annual budget of the colonial government. The market values of government-owned factory output for 1910–1922 are approximated by the total revenues from the state monopolies and the miscellaneous category of the government budgets (see Table A-4). From 1929 on, we have a better basis of estimation. The official source for 1936 and 1939 listed "manufactured commodities" by subdivisions and type of ownership (private or government). From these data we derived the percentage share of government-owned factory output in "manufactured commodities" as shown in Table 3.

As a rough estimate of output produced by the government-owned factories, we applied the 1936 percentage shares to the 1929–1935 data and the 1939 percentage to the 1937–1940 data of "manufactured commodities" to obtain the estimates by the subdivision of the manufacturing sector. These estimates are added to the appropriate subdivision of factory product in the official data.

As for the market values of household-industry output, official data are available only for the years 1931–1939 (Table A-5). The market values for the missing years may be estimated as the difference between "manufactured commodities" and factory output, provided that the factory output data are adjusted to have the same industrial coverage as the "manufactured commodity" data.[14]

Net Value Added

In the absence of any information concerning the cost of

Table 3

PERCENTAGE OF OUTPUT SHARE OF GOVERNMENT FACTORIES IN TOTAL "MANUFACTURED COMMODITIES"

	1936	1939
Textiles	0.30	0.20
Metals	0.00	0.10
Machines, Tools	16.20	2.40
Ceramics	0.95	0.90
Chemicals	0.73	0.60
Wood products	4.99	8.40
Printing	5.37	5.10
Foods	0.03	-
Miscellaneous	50.30	42.90

Source: *Chōsen keizai nempō*, 1939.

intermediate goods in manufacturing, we propose to use Japan's estimates of net product ratios. The Cabinet Bureau of Statistics in Japan estimated the net product ratios for ten subdivisions of the manufacturing sector for 1933, as shown in Table 4. The Japanese subdivisions were used in the official Korean data after 1929. For earlier years, we must reclassify the Korean factory output as shown in Table A-6. The rice-mill output in the Korean data is listed separately from the subdivision of food and beverages,[15] because its net product ratio is extremely low compared to the rest of this subdivision. The average net product ratios for rice mills in 1933 was about 6 percent, as shown in Table 5.

In connection with the use of Japan's net product ratios, one may object to the implicit assumption of similarity between the two countries with respect to the relative magnitude of total costs in the market values of factory output. The assumption may have some validity, however, on the ground that complete economic integration between the two countries was achieved during the early part of the colonial period, and that Japanese capital, technology, and entrepreneurs monopolized this sector. Another objection to the use of these net product ratios may stem from the

Table 4

NET PRODUCT RATIOS OF FACTORY OUTPUT ACCORDING TO JAPANESE SUBDIVISIONS, 1930

(in million K¥)

	Gross Value of Production	Costs			Net Product (Value Added)	Net Product Ratio
		Materials	Fuels & Others	Depreciation		
Textiles	243.0	174.1	8.7	17.0	43.2	17.78%
Metal & Metal products	527.8	397.0	7.1	13.5	110.3	20.90
Machinery	684.6	308.8	26.7	23.0	326.1	47.63
Ceramics	84.6	22.8	6.9	2.9	52.0	61.44
Chemicals	697.5	304.8	81.2	32.8	278.6	39.95
Wood & Wood products	124.8	73.7	5.5	4.3	41.3	33.09
Printing & Bookbinding	44.2	17.7	1.0	0.9	24.7	55.86
Food & Beverages	287.1	176.4	20.4	9.8	80.5	28.04
Miscellaneous	178.1	114.2	4.4	2.4	57.1	32.05
Gas & Electricity	959.0	208.9	175.6	96.3	478.2	49.86

Source: Ohkawa, p.87.

Table 5

REVENUE AND COST OF RICE MILLS, 1933

(per one *suk* of cleaned rice)

	Large Mill[a]	Medium-size Mill[b]	Small Mill[c]
Revenue	21.88	21.90	21.79
Cost	19.80	19.88	19.81
Raw Material			
Power	.15	.14	.11
Others (packaging, etc.)	.53	.55	.59
Total Cost	20.48	20.57	20.51
Net Product Ratios	6.39%	6.08%	9.08%

Source: Chōji Hishimoto, *Chōsen-mai no kenkyū*, Tokyo, 1938.

a. Production of cleaned rice over 500 *suk* per day.

b. Production of cleaned rice between 200–500 *suk* per day.

c. Production of 80 *suk* of cleaned rice per day.

adoption of the 1930 ratios for the entire period under review. "Since the Japanese economy in 1930 was in the midst of a great depression, the figures of net income ratio of this year should be judged as somewhat lower than those in normal years."[16] However, it was also demonstrated that the application of the estimates of 1930 to other years in Japan showed at least the long-term trends in the growth of the manufacturing sector.[17] In short, in spite of its crudeness and limitations, the above method of deriving the net factory output seems to be appropriate for our purpose of measuring long-term trends in the manufacturing sector.

As for the household-industry output, we did not attempt to aggregate individual items into the same subdivisions as the factory output owing to the lack of relevant data. Instead, a single net product ratio was estimated for the entire household-industry output.

From Table 5 (Japan's net product ratios of factory output), we estimated the net product ratios of household-industry output by taking as the cost of production the cost of raw materials in each subdivision. Our assumption in this calculation is that the household-industries in Korea did not rely on the use of machines for production. The net product ratios of the subdivisions were combined into a single ratio, weighted by the percentage distribution of household-industry output in 1935, as shown in Table 6. The net product ratio from this calculation amounted to 40 percent. Our estimates of market and net values for manufactured goods are shown in Table A-7.

Table 6

COMPOSITION AND NET PRODUCT RATIO OF HOUSEHOLD-INDUSTRY OUTPUT, 1935

(in million K¥)

	Gross Value[a]	Percentage Distribution	Percentage of Material Cost in Gross Value of Factory Output[b]	Weighted Net Product Ratio
Textiles	25.8	12.9	28.4	3.8
Metals	5.5	2.7	24.8	.7
Machines & Tools	3.2	1.6	54.9	.9
Ceramics	2.4	1.2	73.0	.9
Chemicals	28.4	14.2	50.3	8.4
Woods	5.0	2.5	41.0	.2
Foods	92.9	46.4	38.6	18.0
Others	37.0	18.5	35.9	6.8
Total	200.2	100.0		39.7

Sources: a. Keijo Shōkō Kaigi-sho, *Chōsen ni okeru gatenkogyo chosa.*

b. Calculated from Table 4.

Fishery and Mining

We find a complete lack of relevant data with respect to the total cost involved in the commodity-product originating in the fishing and mining sectors. Under the circumstances, the estimates of net product ratios had to rely entirely on the appropriate estimates for the Japanese economy.

The market values of fishery products are taken from the official sources without any adjustments. In order to estimate net value added, fishery products are divided into marine products, cultured products, and processed products. The net product ratios of these subsectors are taken from Yamata's estimates of Japanese fishery products with the following modification. In marine products and cultured products, where the bulk of work was done by manual labor rather than by the application of modern technology during the colonial period, the Korean products must have been more labor intensive than the Japanese products during the same period. This observation suggests that the relative cost in terms of payments to other sectors was likely to be lower in Korea than in Japan. In light of this consideration, Yamada's estimates are revised so that the net product ratio is 70 percent for marine products, 60 percent for cultured products, and 35 percent for processed products.[18] Our estimates for the net fishery products are shown in Table A-8.

Mining products consist of nineteen minerals in the statistical yearbooks of the colonial government. In order to derive net output, the net product ratio of 80 percent taken from Yamada's estimates,[19] is applied to the market values. Table A-9 presents the market and net values of mining products.

Price Deflators

Since our primary purpose in this study is the measurement of long-term growth in the commodity-producing sectors, our estimates of net output must be valued according to constant market prices. Two problems have to be dealt with at the outset: the selection of a base year, and the selection of proper price indices as deflators.[20]

The selection of a base year in this study is dictated solely by the nature of the available data. The only price index available for the entire period is the Seoul price index compiled by the Bank of Korea.[21] Beginning in 1939, the Bank of Korea compiled a new series of Seoul wholesale price indices, covering eighty commodities in ten groups, with 1936 as the base year.[22] In order to use these indices, we propose first to construct price indices for agricultural products and for manufactured products covering 1910–1938, and to use the Seoul wholesale price index of the Bank of Korea for the remaining years.

Derivation of Price Indices

Let us begin with the price indices of agricultural products. In order to produce a Paasche index, we first construct volume indices for the five major crops: rice, summer grains, beans, barley, and cotton. Then, the five indices are combined into one index according to the proper weights based on their market values.[23] From the total volume index, PoQn and PnQn/PoQn may be derived, where the subscript "o" denotes the base year, and "n" a given year. The index thus obtained is taken to represent the price index for agricultural product as a whole. The volume indices used for the calculation are presented in Table A-10.

Turning next to the construction of a price index for manufactured products, we encounter serious difficulties in making a consistent volume index for the entire period under review. Apart from the complications arising from the numerous items listed in the official data on manufactured products, the listings are neither complete nor consistent enough to yield the necessary volume index. Thus the following method was devised to utilize existing data on the prices of selected products.

As mentioned, price indices of thirty selected products are available from 1910. Out of these products, ten may be considered to be typical manufactured products in the Korean economy.[24] The ten commodities are grouped according to the subdivision of manufactured products used in Appendix A; Textiles (two products), Chemicals (two products), Foods (five products), and a "Miscellaneous" category (one product). The simple arithmetic averages of the price indices of products within each group are calculated to represent the indices for each subdivision. Finally, the four price indices of the subsectors are combined into one index, with their weights determined by the market values of "manufactured commodities" according to the subdivision.[25]

As for the price index of 1939 and thereafter, the Seoul wholesale price index for grains is used for agricultural product. To compute the price index of manufactured product, the following groups are selected from the Seoul wholesale price index: textile products, metals and metal products, chemicals and fertilizer, and food products. Then their price indices are combined into one index with the relative weights determined by the market values of "manufactured commodities" of 1940.

We can now compare the two estimated indices with the Seoul price index of the Bank of Korea. All three indices are shown in Figure 1 and Table A-11. In general, the three indices show a similar trend over the entire period under consideration. The agricultural price level remained lower than that for manufacturing during most of the period, reflecting the relative price movements between exports (mostly agriculture) and imports (mostly manufactured goods).

Price Indices as Deflators

In order to compute the net products at constant prices, we applied the price index of agriculture to deflate agricultural, forestry, and fishery products. The price index of manufactured products is used to deflate manufactured and mining products. The resulting net products at constant prices are shown in Table A-12.

A detailed analysis of the major trends revealed in these figures is presented in the next chapter. However, it may be useful here to make a general observation on the behavior of net value added in constant prices. As may be expected, the agricultural products show a substantial degree of annual fluctuation due to the annual changes in the total amount of crops harvested. The determining factor here was the volume of rice production. For example, the sudden increase in the aggregate value of the agricultural product during 1929–1930, and 1936–1937 reflected the

FIGURE 1. PRICE INDICES, KOREA, 1910-1945
(1936=100)

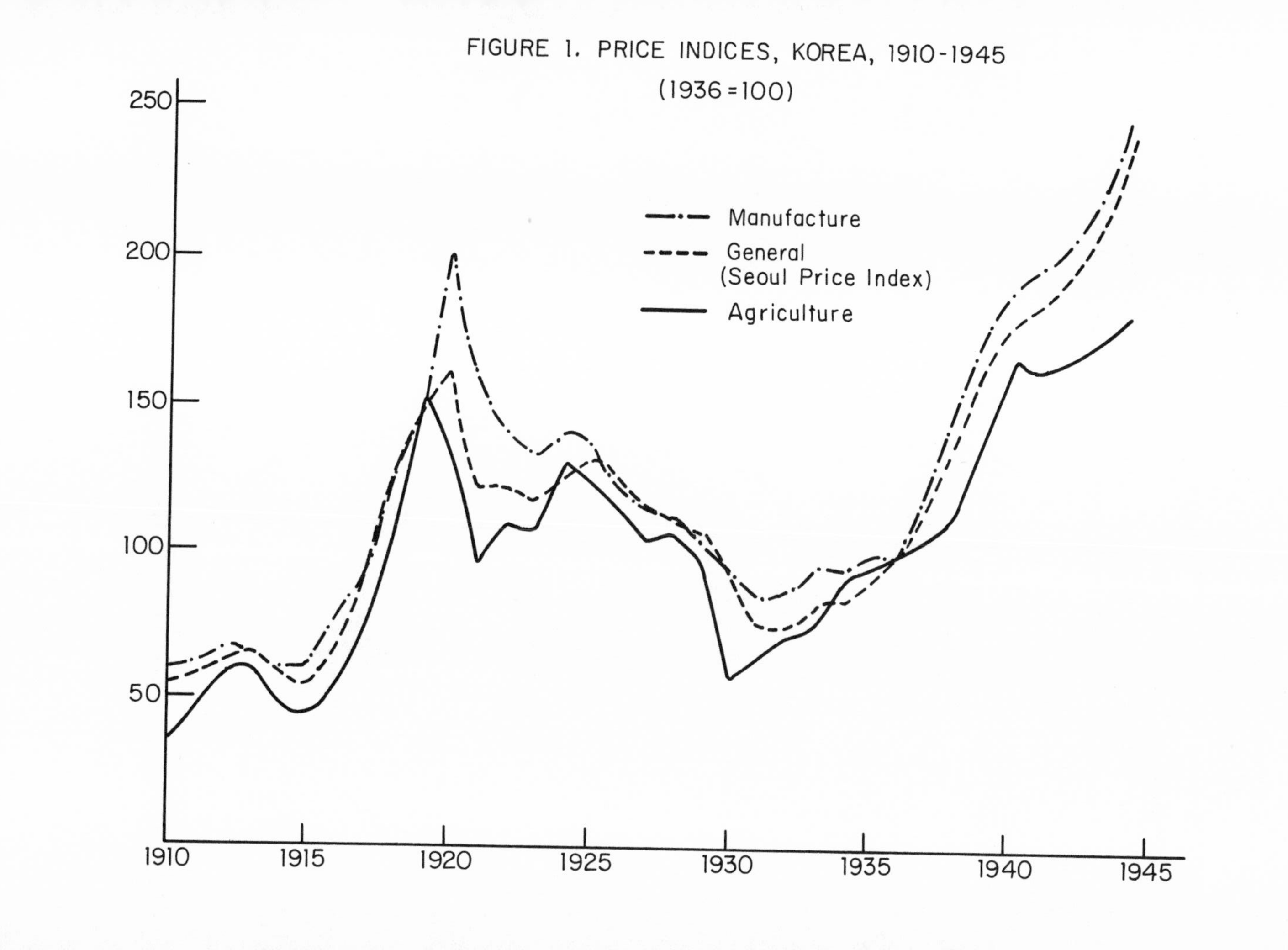

extremely good harvests of rice in 1930 and 1937. The opposite movement during 1938–1939 should be attributed to the bad harvest of rice in 1939. The estimates of non-agricultural products show an upward trend during the entire period under review, particularly in mining and manufactured products during the latter years of the period, reflecting the rapid industrialization of the 1930s.

Chapter III

MEASUREMENT OF OVER-ALL GROWTH

Initial Conditions

A brief examination of economic conditions at the time of annexation may provide a useful background for the understanding of economic changes during the colonial period. Let us first examine the industrial distribution of the labor force. This may be approximated by the industrial distribution of households in 1910, as compiled by the colonial administration. The relevant data are summarized in Table 7. According to the figures in Column 1, over 84 percent of all households were engaged in agriculture while fewer than one percent of the households were in the manufacturing sector. Granted that the data do not show any industrial distribution of workers other than the head of each household, they nevertheless reveal that the national economy was dominated by the agricultural sector during that period. The extremely low labor share of the manufacturing sector indicates that the major portion of non-agricultural products consumed domestically was provided by members of each household, with little division of labor in the economy as a whole.

The predominance of the agricultural sector in the economy may be seen from another angle by examining the characteristics of cities[1] at the time of Korea's annexation. According to Columns (2) and (3) of Table 7, the industrial distribution of employment for cities, other than Seoul, is not basically different from the national pattern (Column 1): the agricultural sector predominates, followed by commerce and other service sectors. Thus, it is apparent that the mode of living in Korean cities, except Seoul, did not fundamentally differ from that of the rural areas. These data demonstrate that there was no significant development of urbanization and division of labor in the Korean economy at the time of annexation in 1910.

Since Japanese residents occupied strategic positions in the

Table 7

INDUSTRIAL DISTRIBUTION OF KOREAN HOUSEHOLDS
(May 1, 1910)

Industrial Origin	Total (1)	Seoul (2)	Cities (except Seoul) (3)
	Absolute Figures		
Agriculture (including Forestry)	2,433,450	8,643	92,911
Fishery	33,646	87	3,987
Mining	1,429	60	40
Manufacturing	22,943	3,310	1,809
Commerce	178,780	13,672	12,760
Civil Service	15,758	2,095	612
Others[a]	177,647	15,257	11,609
Unemployed	31,123	12,886	2,950
Total	2,894,776	56,010	126,678
	Percentage Shares		
Agriculture (including Forestry)	84.10	15.43	73.34
Fishery	1.16	0.16	3.15
Mining	0.02	0.10	0.04
Manufacturing	0.79	5.91	1.43
Commerce	6.18	24.41	10.07
Civil Service	0.54	3.74	0.48
Others	6.13	27.24	9.16
Unemployed	1.08	23.01	2.33
Total	100.00	100.00	100.00

Source: Chōsen Sōtokufu, *Chōsen tōkei nempō*, 1909, pp. 63-79.

a. The entries represent all the remaining occupations, including such social classes as *yanban, yooga* of upper strata.

Korean economy during the transitional period, our analysis must include their industrial distribution in order to assess the over-all pattern of employment in the economy. For this purpose, data on the industrial distribution of Japanese households in Korea are shown in Table 8, along with the combined figures for Korean and Japanese households. The figures accord with our expectation that Japanese immigrants in Korea engaged mainly in commerce, civil service, manufacturing, and in other professional occupations. However, the industrial distribution of total households, including both Koreans and Japanese, reveals once again the predominance of agriculture in the Korean economy.

We may now compare the Korean economy of 1910 with the Japanese economy during the latter's transition to modern economic growth.[2] Let us begin with the industrial distribution of workers in the two countries. It should be noted here that only a crude comparison can be made because there are insufficient data for Korean workers; they have been estimated from the industrial distribution of households. The comparison, as shown in Table 9, seems to indicate a resemblance between the industrial structures of the two countries: both economies were dominated by the agricultural sector, followed by services. The large share in the Korean service sector may be attributable to the rapid expansion of commerce and international trade between the open-door policy of 1876 and the annexation of 1910.

The comparison of industrial composition of product has to be limited to the commodity-producing sectors because the relevant data for services are non-existent for Korea. According to our estimates of net commodity-product,[3] over 91 percent of the total originated in agriculture (including forestry and fishery) and the remaining 9 percent in the manufacturing sector during 1910–1912. The corresponding percentage distributions for Japan during 1878–1880 were 88 percent and 12 percent respectively.[4] Insufficient data prevent us from assessing the difference in the levels of aggregate products.

In view of the above similarities in the industrial distribution of employment and commodity-product between the two countries, it may be argued that the Korean economy at the beginning of the colonial period corresponded roughly to the Japanese economy of the 1870s.

Table 8

INDUSTRIAL DISTRIBUTION OF EMPLOYMENT
(December 1910)

Industrial Origin	Japanese Households (1)	Total Households (Japanese and Koreans) (2)
	Absolute Figures	
Agriculture (including Forestry)	2,210	2,435,660
Fishery	1,423	35,069
Mining	--	1,429
Manufacturing	5,619	28,562
Commerce	14,568	193,348
Civil Service	8,724	24,582
Others	16,730[a]	194,500
Unemployed	1,718[b]	32,841
Total	50,992	2,945,991
	Percentages Shares	
Agriculture (including Forestry)	4.33	82.68
Fishery	2.79	1.19
Mining	--	0.01
Manufacturing	11.02	0.97
Commerce	28.56	6.56
Civil Service	17.11	0.83
Others	32.75	6.65
Unemployed	3.44	1.11
Total	100.00	100.00

Sources, by column:

(1) Chōsen Sōtokufu, *Chōsen tōkei nempō*, 1910, pp. 81-87.
(2) Table 1.

a. Includes professionals.
b. Includes the households that were not yet permanently settled.

Growth of Commodity-Product

The scale of over-all growth must be measured according to the net value added estimated in Chapter II. In Table 10, the rele-

Table 9
PERCENTAGE DISTRIBUTION OF EMPLOYMENT
IN JAPAN & KOREA

	Japan 1878–1880 (1)	Korea 1910 (2)
Agriculture	83%	84%
Manufacturing	5%	1%
Services	12%	15%

Sources by column:

(1) Henry Rosovsky, "Japan's Transition to Modern Economic Growth," p. 93.
(2) From Table 7.

vant data by industrial origin are summarized for the selected years. The figures in this table represent the annual averages of five-year totals to smooth out annual variations from the primary data. Their average annual growth rates are given in Table 11.

The data show a sustained and substantial growth in total commodity-product during the period. Within the general upward trend, we find very high growth rates in total commodity-product during both the early and later years. The high growth rate of the 1920s was at least partly caused by an upward bias in the data resulting from enlarged statistical coverages as the colonial government became more effective in compiling data after the annexation of 1910. Even though we can't measure the exact magnitude of the upward bias in the data, it seems evident that the rate of growth during the first decade after annexation cannot be fully or perhaps even generously discounted on the basis of the upward bias of the data alone. As already mentioned, various policy measures were used during the first decade of the colonial period in order to expand agriculture in Korea, and this was done by mobilizing the substantial amount of "slack" in the rural economy at the time of annexation.

Table 11 shows the rapid growth in agriculture and forestry during the second phase of the colonial period, reflecting the effects of government policies designed to expand agriculture,

Table 10

COMMODITY-PRODUCT BY INDUSTRIAL ORIGIN

GROSS VALUE ADDED

(in billion K¥, 1936 prices)

Annual Average	A-sector	M-sector	Total
1910-1915	0.73	0.04	0.77
1916-1921	0.88	0.06	0.94
1922-1927	0.95	0.10	1.05
1928-1933	1.11	0.16	1.27
1934-1939	1.18	0.37	1.55
1940	1.16	0.49	1.65

Source: Computed from Table A-12.
A-sector: Agriculture, Forestry, Fishery
M-sector: Mining, Manufacturing

Table 11

ANNUAL GROWTH RATES OF COMMODITY-PRODUCT

BY INDUSTRIAL ORIGIN

(percentages)

Period	A-Sector	M-Sector	Total
1910/15–1916/21	3.2	7.0	3.4
1916/21–1922/27	1.3	8.9	1.9
1922/27–1928/33	2.7	8.2	3.2
1928/33–1934/39	1.0	15.0	3.4
1910/15–1934/39	2.0	9.7	3.0

Source: Computed from Table 10.

particularly rice production.

The complexion of the national economy was drastically changed during the third phase of the colonial period by the conspicuously high growth rates in manufacturing and mining as shown in Table 11. The dominant sector of the economy shifted from agriculture to manufacturing. For example, the contribution rate of the manufacturing sector to the total commodity-product growth rose from 14.4 percent in 1910–1920 to 56.8 percent in 1930–1940, as shown in Table 19.

Population and Per Capita Product

At the outset, a word of explanation is in order concerning the nature of population data in Korea during the colonial period. After Korea's annexation in 1910, a system of registering members of each household was introduced, which in turn provided the basis for official estimates of the total population at the end of each year. In addition, the first nationwide population census was taken in 1925, and was repeated thereafter every five years, with the last census taken in 1944.[5] Thus two types of population data are available for the period under review. However, because the census data are generally believed to be more accurate than the official year-end estimates,[6] we used the census data in the analysis for the entire period, except for the early years for which census data did not exist. For these years we made some modifications of the year-end estimates in order to eliminate the obvious bias in the data.[7] Table 12 summarizes the magnitude of population growth in Korea during 1910–1944. By using the population data of Table 12 as bench marks, we estimated the annual population for the entire period as shown in Table 13.

During the period observed in this table, the total population increased by about 70 percent with an average annual growth rate of 1.6 percent. Since the figures in Table 13 represent the Koreans only, its growth must have come largely from natural increase, that is, the difference between birth and death rates.[8] The accuracy of official data based on the household registration was seriously impaired because the system of registering members of households was not yet an accepted social routine in Korea. For this reason, the U.N. estimates of vital statistics of Korea for 1920–1940 are used in Table 14.[9]

According to these figures, death rates show a definite decline after 1930, whereas birth rates seem to have fluctuated at a high level. Unforturately, there are no reliable vital statistics for the years prior to those covered by the United Nations estimates.

Table 12

KOREAN POPULATION, 1910–1944

Year	Population (thousands of persons)	Net Increase	Percentage Increase
1. 1910	14,766	–	–
2. 1915	15,958	1,192	8.0
3. 1920	17,264	1,306	8.3
4. 1925	19,020	1,756	10.1
5. 1930	20,438	1,418	7.4
6. 1935	22,208	1,770	8.6
7. 1940	23,547	1,339	6.0
8. 1944	25,133	1,586	6.7

Sources by line:

1. See note 7, Chapter III.

2-3. *Tōkei nempō,* 1915, 1920.

4-8. Chōsen Sōtokufu, *Chōsen kokusei chosa hokoku,* 1925, 1930, 1935, 1940, 1944.

However, in view of the rapid growth of total population during the early years of the colonial period, it is quite evident that death rates must have reduced substantially as the country moved from the transitional period to the colonial period. In short, the rapid population growth in Korea during this period may be explained in terms of a substantial and continuous decline in death rates, while birth rates remained at a high level.[10]

The decline in death rates was largely brought about by the restoration of social order and improved public health measures. On the other hand, the colonial policies designed for the rapid growth of population[11] produced a high birth rate throughout the colonial period. Thus the demographic trends of the colonial period in Korea did not reach a mature stage where, after a rapid population expansion, there is a decline because of declining birth rates in the process of sustained economic growth.

Of special interest in Table 14 is the trend of emigration. There is a striking increase during the latter part of the period, lessening the population pressure in Korea. That rising trend was created by

Table 13
ESTIMATES OF ANNUAL POPULATION
(thousands of persons)

Year	Population	Year	Population
1910	14,766	1928	19,889
1911	15,002	1929	20,187
1912	15,242		
1913	15,486	1930	20,438
1914	15,743	1931	20,785
1915	15,958	1932	21,138
1916	16,213	1933	21,497
1917	16,472	1934	21,862
1918	16,736	1935	22,208
1919	17,004	1936	22,474
		1937	22,744
1920	17,264	1938	23,017
1921	17,609	1939	23,293
1922	17,961		
1923	18,320	1940	23,547
1924	18,686	1941	23,924
1925	19,020	1942	24,307
1926	19,305	1943	24,696
1927	19,595	1944	25,133

Note: With the population data of Table 12 as bench marks, the annual data are estimated by applying the annual average growth rates between two bench-mark years.

the increasing migration of Koreans during the 1930s. The volume of migration during this time cannot be precisely measured because records of the annual figures were not maintained. However, its magnitude may be observed roughly from the records of the Korean residents in Manchuria and Japan. The relevant data show that Korean residents in Manchuria increased from 600,000 to 1.4 million during 1920–1940. Korean residents in Japan rose from 41,000 to 419,000 during 1920–1930, and the figure reached 1.2 million in 1940.[12] Thus, the decline in the population growth rates during the later part of the colonial period was not caused by

Table 14

VITAL STATISTICS OF KOREA, 1920–1940
(annual averages per 1,000 population)

Period	Birth Rate	Death Rate	Natural Increase	Emigration Rate	Actual Increase
1920–1925	43.4	19.4	24.0	0.0	24.0
1925–1930	39.0	21.5	17.5	- 3.0	14.5
1930–1935	38.7	18.9	19.8	- 3.0	16.8
1935–1940	42.0	18.7	23.3	-10.0	13.3

Source: The United Nations, Population Division, *Future Population Estimates by Sex and Age: The Population of Asia and the Far East, 1950–1960*, (Report IV), p. 49.

any reduction in the natural rate of increase.

The rapid growth of total population, along with the industrialization of the 1930s, led to drastic changes in the distribution of population between rural and urban areas.[13] These are clearly reflected in Table 15, which indicates percentage increase of population in urban and rural areas. During 1925–1940, the share of urban population rose from 3.3 to 13.1 percent.

Let us now observe the growth of population in relation to that of the commodity-product. For this purpose, the average annual growth rates of total population are compared with the growth rates of the per capita commodity-product during the colonial period in Table 16. The figures indicate that growth rates of per capita commodity-product followed a course similar to that of the total commodity-product. Growth rates were relatively high during the early and later years of the colonial period. In general, the growth rates of the per capita commodity-product were higher than those of the total population, except during 1915–1925. For the period as a whole, the level of per capita commodity-product rose by over 80 percent. These findings once again demonstrate that the colonial period witnessed substantial economic growth of commodity-product, surpassing by far the rapid growth of total population.

Table 15

PERCENTAGE INCREASE IN URBAN AND RURAL POPULATION 1925-1940

Population	Periods		
	1925-1929	1930-1934	1935-1939
Total Population (Koreans and Japanese)			
Urban	40.0	35.0	75.7
Rural	6.4	7.2	1.0
Koreans only			
Urban	46.4	40.0	91.0
Rural	6.2	7.2	1.0
Japanese only			
Urban	21.3	24.6	30.2
Rural	16.7	10.0	4.6

Source: Taeuber and Barclay, "Korea and the Koreans in the Northeast Asian Region," *Population Index,* xvi.4 (October 1950), 284.

Table 16

AVERAGE ANNUAL GROWTH RATES OF TOTAL POPULATION AND PER CAPITA COMMODITY-PRODUCT 1910-1940 (percentages)

Period	Population	Per Capita Commodity-Product
1910-1920	1.6	2.0
1915-1925	1.8	0.3
1920-1930	1.5	1.7
1925-1935	1.1	2.3
1930-1940	1.4	3.1

Sources: Tables 12 and A-12.

Chapter IV

STRUCTURAL CHANGES

This examination of structural changes will supplement the preceding analysis of the general characteristics of the economy during the colonial period. We begin with structural changes in production and the distribution of the labor force. This is followed by an analysis of structural changes in the use of total available products, based on commodity-flow analysis.

Structure of Commodity-Product

Table 17 shows the percentage distribution of net commodity-product by industrial origin, based on current prices. The striking feature of the table is the magnitude and speed of the changes in the industrial composition of the total commodity-product. During the three decades under review, the share of the agricultural product declined from 92.5 to 59.6 percent, while the proportion of manufactured product rose from 3.6 to 23.9 percent. On the whole, the data indicate a remarkable speed of industrialization, and the short span of time involved in the industrialization should be kept in mind as one of the peculiarities found in the Korean economy during the colonial period.

For the study of long-term trends, it is important to eliminate the effects of relative price changes in the industrial sectors. Ideally, we should adjust the current values for price changes in each sector.[1] However, such elaborate calculations are not feasible with the present data. Thus we use the price index of total commodity output to derive the industrial composition of commodity-product at constant prices. The results are presented in Table 18.

These figures reveal a roughly similar pattern of structural changes during the colonial period, as shown in Table 17 based on current values. The share of the agriculture sector progressively declined, while that of the industrial sector showed a substantial increase. It may be noted also in this table that the relative share of

Table 17

PERCENTAGE DISTRIBUTION OF NET COMMODITY-PRODUCT BY INDUSTRIAL ORIGIN, 1910-1940

(based on current values)

Annual Average	Agriculture & Forestry	Fishery	Mining	Manufac-turing	Total
1910-1915	92.5	2.3	1.6	3.6	100.0
1916-1921	87.9	3.1	1.7	7.3	100.0
1922-1927	85.0	4.1	1.3	9.6	100.0
1928-1933	79.5	5.2	2.3	13.0	100.0
1934-1939	67.8	5.8	6.4	20.0	100.0
1940	59.6	7.7	8.8	23.9	100.0

Sources: Computed from Tables A-3, A-7, A-8, A-9

Table 18

PERCENTAGE DISTRIBUTION OF NET COMMODITY-PRODUCT BY INDUSTRIAL ORIGIN, 1910-1940

(based on 1936 prices)

Annual Average	Agriculture & Forestry	Fishery	Mining	Manufac-turing	Total
1910-1915	92.9	2.3	1.3	3.5	100.0
1916-1921	89.9	3.2	1.4	5.5	100.0
1922-1927	86.0	4.2	1.2	8.6	100.0
1928-1933	82.1	5.3	2.0	10.6	100.0
1934-1939	70.3	5.6	5.8	18.3	100.0
1940	61.4	8.3	8.3	22.0	100.0

Source: Computed from Table A-12.

the industrial sector was always lower in terms of constant prices than the share based on current values. This was caused by a discrepancy in relative price changes between the two sectors. The rate of change in the general price level of agricultural products was lower than that of the manufactured product during most of the colonial period. Nevertheless, the figures in both tables confirm the conclusion that Korean economic growth during the colonial period led to a substantial change in the industrial composition of output produced in favor of the industrial sector, particularly during the 1930s.

The differential growth rates among the commodity-producing sectors and the resulting changes in the industrial composition of total output produced some basic changes in the relative contribution of each sector to the over-all growth of commodity-product. This may be observed in the contribution rates according to industrial sectors, as presented in Table 19.[2]

According to these figures, 85.6 percent of the over-all growth during the 1910s was attributable to the expansion of the A-sector and only 14.4 percent to the growth of the M-sector. Agriculture was clearly the leading sector of Korean economic growth during this time. However, the rapid industrialization of the 1930s shifted the leading role from agriculture to manufacturing. During the 1930s, for example, the contribution rate of the M-sector was 79.1 percent, whereas the A-sector accounted for only 20.9 percent of the over-all growth.

Structure of Employment

At the outset, the concept of "gainful worker" in the official data of employment needs to be clarified. This concept is exactly the same as that used in the Japanese data, and has a meaning quite different from the usual definition of labor force. It refers to:

> the status of the individual in question. There is no precise definition about how much work a person has to do in order to be counted as gainfully occupied. In practice, people who work as little as one hour per week may be included.[3]

Thus, the available data fail to show any change in the total amount of the actual labor force in the economy, and we are unable to distinguish between part-time and full-time workers. In addition to the above limitations, during the entire colonial period a census of gainful workers was taken only in 1930 and 1940.[4] Accordingly,

Table 19

CONTRIBUTION RATES BY INDUSTRIAL ORIGIN
(percentages)

Period	A-sector	M-sector	Total
1910/15–1916/21	85.6	14.4	100.0
1916/21–1922/27	63.4	36.6	100.0
1922/27–1928/33	73.3	26.7	100.0
1928/33–1934/39	20.9	79.1	100.0

Sources: Computed from Tables 2 and 3. See note 2 (Chapter II) for explanation.

reliable data on gainful workers are not available for the other years.

Growth and Industrial Distribution of Workers

Table 20 presents the number of gainful workers by industrial origin. The data for 1910 are official estimates of households by industrial origin, whereas the data for 1930 and 1940 are census data. Even though the data used in the table differ in reliability and in methods of compilation, broad aspects of the employment pattern during the colonial period can be discerned. Data for the industrial distribution of workers until 1930 agree with our expectation that the agricultural sector employed the major portion of the total labor force, ranging from 84.1 percent in 1910, to 78.4 percent in 1930.

As for the decade of the 1930s, we find a substantial reduction of workers in Table 20. However, these data may obscure the true picture of the total labor force because they fail to show the change in composition between full-time and part-time, or between male and female workers. To obtain further insight into the total use of the labor force during this decade, the data in Table 20 are divided in terms of male and female workers, as shown in Table 21. According to this table, female workers declined by 712,000 in contrast to a net increase of 215,000 male workers during 1930–1940. With the limited data at hand, it is impossible to meas-

Table 20
DISTRIBUTION OF "GAINFUL WORKERS"
BY TYPE OF OCCUPATION, 1910, 1930, 1940
(thousands of workers)

	1910[a] Number of Workers	1930	1940	Differences 1930–1940
Total	2,914	9,699	9,195	-504
Agriculture	2,436	7,652	6,685	-967
Fishery	35	122	135	13
Mining	1	32	176	144
Manufacturing	29	566	495	-71
Others	413	1,327	1,704	377

Sources: 1910—year-end estimates, taken from the statistical yearbook (1910) of the Government-General.
1930, 1940—Population Census Data.

a. Number of households

sure accurately the net change in the use of the total labor force associated with these figures. However, it is rather common that, at a time of rapid industrialization, male workers in general tend to work longer hours per day and to have higher productivity than female workers. For this reason, it may be argued that the decline of workers during the decade of the 1930s, as shown in Table 20, was the result of the substitution of a smaller number of "primary workers" for "secondary workers" rather than of an absolute decline in the total labor force.[5]

The magnitude of reduction in the relative share of secondary workers may be seen from another angle. According to the figures in Table 20 and the population census data, the crude activity rate[6] for the population as a whole declined from 46 to 38 percent during the period 1930–1940. The crude activity rate of female workers declined from 32 to 22 percent and that of male workers from 59 to 54 percent.

The substantial reduction of secondary workers for the economy

Table 21

DISTRIBUTION OF "GAINFUL WORKERS" BY OCCUPATION AND SEX, 1930, 1940
(thousands of workers)

	Male Workers			Female Workers		
	1930	1940	Difference	1930	1940	Difference
Agriculture	5,032	4,554	-478	2,620	2,131	-489
Fishery	100	119	19	22	16	6
Mining	31	170	139	1	6	5
Manufacturing	284	439	155	282	56	-226
Others	923	1,303	380	398	402	4
Total Workers	6,370	6,585	215	3,343	2,611	-732

Source: Population Census Data, 1930, 1940.

as a whole reflects some basic changes in the pattern of employment during the 1930s. The decade marked the rapid expansion of heavy industries and mining, all related to Japan's active preparation for the World War II. The migration of Korean workers (voluntary and forced) was prevalent during this period. The impact of these changes on labor distribution was clear; the agricultural sector showed a substantial decline in the absolute number of workers, whereas the number of male workers increased in other sectors. The drastic change in the structure of manufacturing from handicraft to heavy industries was accompanied by the substantial increase of male workers replacing female workers.

Productivity Changes Per Worker

In view of the declining activity rates during the 1930s, it seems evident that the substantial growth of per capita commodity-product during the 1930s, as observed in the preceding chapter, must have been the result of a considerable increase in productivity per worker rather than of an expansion in the labor force, as was the case before 1930. Thus, our next task is to examine the major factors that brought increasing productivity per worker during the decade of the 1930s. In general, higher productivity per

Table 22

PERCENTAGE DISTRIBUTION OF "GAINFUL WORKERS" BY INDUSTRIAL ORIGIN, 1930–1940

	1930			1940		
	Total	Male	Female	Total	Male	Female
Total Employment	100.0	100.0	100.0	100.0	100.0	100.0
Agriculture	78.4	79.9	78.8	72.7	69.1	81.6
Fishery	1.2	1.6	0.7	1.5	1.8	.6
Mining	.3	.5	0.0	1.9	2.6	.2
Manufacture	6.0	4.5	8.5	5.4	6.7	2.1
Others	14.1	14.5	12.0	18.5	19.8	15.5

Source: Table 21.

worker may be obtained by the following methods: the inter-industry shift of workers from low-productivity to high-productivity sectors, increased capital intensity per worker, and improved quality of the labor force.

Table 22 presents the percentage distribution of workers by sex and industrial origin. The inter-industry shifts of gainful workers show different trends between male and female. While the distribution pattern of female workers shifted from manufacturing to the agriculture and service sectors, that of male workers shifted from agriculture to the manufacturing and mining sectors. To determine the net effect of these changes on the over-all growth of the commodity-product in the economy, we must begin by examining the differing productivity per worker according to industrial origin.

Since the available data are not adequate to measure the productivity per worker by industrial origin, we used net value added (NVA) per worker in the following analysis. Table 23 presents NVA per worker in the commodity-producing sectors.

The figures in this table agree generally with our expectations: the product per worker in agriculture was lower than in non-agricultural sectors, and the product per worker as a whole increased substantially during the industrialization of 1930–1940. High pro-

Table 23

COMMODITY-PRODUCT PER WORKER BY INDUSTRIAL ORIGIN, 1930–1940

(annual averages 1921–1931 and 1939–1941 at 1936 prices)

	1930	1940
Country-wide average product per worker[a] (K¥)	165.7	221.7
Agriculture, Forestry & Fishery		
Product per worker (K¥)	160.1	170.5
Relation to country-wide average	0.9	0.8
Mining		
Product per worker (K¥)	656.3	761.4
Relation to country-wide average	4.0	3.4
Manufacturing		
Product per worker (K¥)	213.8	735.4
Relation to country-wide average	1.3	3.3

Source: Computed from Tables 20 and A-12.

a. Total commodity-product divided by the total number of "gainful workers."

ductivity per worker in agriculture during this decade was achieved largely by a reduction of workers and the increasing use of commercial fertilizers (intensive farming), as shown in Chapter V.

Productivity increase in non-agricultural sectors was made through rapid industrialization. However, some peculiarities revealed in Table 23 deserve explanation. We note that in 1930 the level of product per worker in the manufacturing sector was not substantially higher than in agriculture, indicating that a large proportion of workers in the manufacturing sector were engaged in handicraft industries. In contrast, productivity per worker in the manufacturing sector from 1930 to 1940 should be attributed to the fundamental changes made in this sector through rapid industrialization, the reduction of female workers, and the shift of male workers from handicraft industries to modern factories.

As for the mining sector, a moderate decline in the relative level of NVA per worker in 1940 was caused by the tremendous increase of gainful workers in this sector (an increase of over 500 percent). The reason for this will be explained later in connection with the exodus of workers from the agricultural sector.

In short, our findings so far indicate that the substantial rise in average productivity per worker was achieved by inter-industry shifts of workers (from agriculture to mining and manufacturing) during the process of rapid industrialization. However, patterns of inter-industry shifts should be examined in further detail because the data used so far cover the total population, including Japanese workers in Korea. By observing the changing patterns of employment in terms of nationality,[7] the "imposed" nature of industrialization in Korea may be illuminated.

We begin with the employment pattern of Japanese workers in Korea. Since census data on gainful workers according to nationality are available only for 1930 and 1940, the occupational distribution of the total number of Japanese in Korea for the earlier years is used in Table 24[8]. The general pattern of employment, revealed in these figures, indicates that Japanese workers in Korea constituted the leading group (economic elite) in the country, with heavy concentration in non-agricultural sectors. Furthermore, the major changes in the employment pattern of Japanese workers during the colonial period corresponded to the similar changes in policy objectives of the colonial administration. For example, during the early years of the colonial period, when the major policy objectives were development of food grain production and establishment of a solid colonial foundation in the country, we find a heavy concentration of Japanese in agriculture, commerce, transportation, and civil services.

Of significance in Table 24 is the employment pattern of Japanese workers associated with the rapid industrialization of the 1930s. The data show a substantial reduction in the relative shares of Japanese workers in the agricultural and miscellaneous categories, while substantial gains are registered in manufacturing, transportation, and communication. The data also indicate that the new pattern of employment in 1940 was created not only by the inter-industry shifts of Japanese workers during the 1930s, but also by

Table 24

OCCUPATIONAL DISTRIBUTION OF JAPANESE WORKERS IN KOREA, 1915, 1930, 1940

(thousands of persons)

	1915[a]	1930[b]			1940[b]			Percentage Distribution of Total Workers		
		Male	Female	Total	Male	Female	Total	1915	1930	1940
Agriculture	11	13	7	20	11	4	15	11	9	5
Fishery	3	7	-	7	6	1	7	3	3	3
Mining	c	1	-	1	9	0	9	-	-	3
Manufacturing	10	37	3	40	65	5	70	10	18	25
Commerce	29	34	25	59	39	25	64	29	26	23
Transportation & Communication	c	19	2	21	40	3	43	c	9	15
Civil Service & Professional	26	66	6	72	60	10	70	26	32	25
Other	21	4	5	19	1	4	5	15	3	1
Total Workers	-	181	48	229	231	52	283	-	100	100

a. Occupational Distribution of Total Japanese in Korea. Computed from Chōsen Kōsei Kyōkai, *Chōsen ni okeru jinkō ni kansuru sho tōkei.*

b. Population Census data.

c. Included in "Commerce."

the substantial increase of new workers from Japan.[9] There was a 24.1 percent increase of Japanese workers during 1930–1940, whereas the total number of Korean workers declined, as shown in Table 20. In short, the 1930s' rapid industrialization in Korea had a far-reaching impact on the employment pattern of Japanese workers.

Table 25

OCCUPATIONAL DISTRIBUTION OF KOREAN MALE WORKERS
1930–1940

	1930		1940		
	Number of Workers (thousands)	Percentage Distribution	Number of Workers (thousands)	Percentage Distribution	Difference
Agriculture	5,019	81.1	4,543	71.5	-476
Fishery	93	1.5	113	1.8	20
Mining	30	0.5	161	2.5	131
Manufacturing	247	4.0	374	5.9	127
Others	801	12.9	1,163	18.3	362
Total	6,190	100.0	6,354	100.0	164

Source: Population Census Data.

The industrialization of the 1930s also gave rise to a changing pattern of distribution for Korean workers. This may be observed in Table 25, which presents the occupational distribution of Korean male workers. First of all, there was only a 2.6 percent increase in Korean male workers during 1930–1940. Thus, the major changes in the employment pattern must have been caused largely by inter-industry shifts of Korean workers. According to the figures, a substantial degree of inter-industry shifts took place during the 1930s from agricultural to non-agricultural sectors. To these shifts of male workers should be added our earlier findings on the reduction of female workers when we consider the major changes in the employment pattern of Korean workers during the period of rapid industrialization.

As for the degree of capital intensity per worker during the rapid industrialization of the 1930s, data are non-existent. However, our preceding analysis has already suggested the general trend. For example, we observed a substantial rise in productivity per worker in agriculture, achieved through new methods of cultivation, requiring a greater amount of capital per worker (for example, fertilizers, better seeds, and so forth). In manufacturing, the very nature of the rapid industrialization called for capital-intensive techniques already mentioned. This trend was also evident in the composition of capital goods. Thus, it seems evident that the average capital per worker in the Korean economy increased substantially during this decade. The rising trend also explains the over-all increase of productivity per worker in most of the industrial sectors in Table 23.

Domestic Use of Commodity-Product

We now deal with the major changes in the domestic use of finished goods. The approach here is similar to the commodity-flow analysis following Professor Kuznets's exposition.[10] We do not attempt, however, to provide an extensive commodity-flow analysis because of the acute shortage of data. Compared to an extensive commodity-flow analysis, the present analysis deals exclusively with the data on commodity-product, excluding services, construction, and government. Second, only a broad breakdown of finished goods by industrial origin (instead of individual goods) is possible with the limited data at hand. Third, the flow of goods to the ultimate consumers has to be estimated by the total of consumers' goods available for domestic use, including inventories, owing to the non-existence of inventory data. Fourth, goods are valued at producers' prices without adjusting for transportation costs and distribution charges to the ultimate users. In spite of these limitations, the present analysis will serve our purpose of observing long-term trends in the domestic use of finished products during the colonial period.

The analysis will also be useful for understanding the welfare implications of Korean economic growth during the period under review. It was shown in the preceding analysis that the substantial rise in domestic production during the colonial period, along with the import surplus in the international flow of goods, represented an even higher level of commodity-product available for domestic use. However, the rising trend does not necessarily mean similar changes in the level of domestic consumption for the majority of inhabitants. This is largely determined by the pattern of commodity distribution between consumption and capital formation, and by the distribution of income.

Classification of Commodities

In order to measure, without duplication, the flow of commodities to consumption and capital formation, it is essential to secure a careful classification of commodities between finished goods and unfinished goods because the flow of commodities includes only the total of finished goods at producer prices. After this classification, all the finished goods must be divided into consumers' goods and producers' goods.[11] The total domestic flow of commodities is measured by the domestic production of finished goods plus the net balance of foreign trade of finished commodities. Thus, the above classification of commodities must be applied not only to domestic production, but also to foreign trade commodities.

When the commodity classification is done in terms of broad breakdowns (rather than by individual commodities), the scope of so-called mixed commodities increases,[12] and some degree of arbitrariness is bound to be used in the classification.[13] Since our judgments made on the classification of commodities have some bearing on the final results, we must begin with a detailed discussion on the classification of commodities used in the present analysis.

Among agricultural products, finished goods include crude foodstuffs (that is, grains, fruits and vegetables) except those used for the manufacturing of food products.[14] Excluded from our classification of finished goods in agriculture are cotton and special crops for industrial uses, cocoon products, fertilizers produced in farm households, and livestock products.[15] Forestry products used for fuel and medicinal herbs comprised finished goods originating in this sector. Excluded are lumber and woods used for the manufacturing of wood products.

Fishery products in official data are listed by the broad categories of marine products, cultured products, and processed products. However, some portion of marine products is used by producers for the processed products. Thus, finished goods in fishery are estimated to be the total of fishery products at producer prices with the portion of marine products used for the processed products subtracted.[16]

Data on finished goods of manufacturing are obtained from the official publications of "manufactured commodities," classified by the subsectors of manufacturing used for the estimates of net values in Chapter II. The general method used here is to separate finished goods in each subsector of manufacturing, and then to classify them into consumers' goods and producers' goods.

Finished goods in textiles include all the products of cotton textiles, woolen textiles, hemp textiles, and mixed textiles. They also include other textile products such as towels, blankets, and knit cloth. Excluded from finished textile goods are all types of yarns used in textile industries. The finished textile goods are classified as consumers' goods.

Among the metal products in official data, finished goods comprise cast metal products (other than machine and tools) and other metal products such as wires, containers, nails, and other building materials.[17] The market values of these products are recorded as producers' goods. Included also in producers' goods are market values of machines and tools in official data.

Consumers' goods (among finished goods) in chemical products include drugs and other pharmaceutical products, soaps and cosmetics, paper, and candles. On the other hand, rubber products, paints for building and construction, and miscellaneous chemicals in official data comprise producers' goods. Excluded from finished goods of chemical products are chemical fertilizers, all types of oil and gasoline,[18] dyes, and semi-finished chemicals for industrial use.

Market values of wood products and foods in official data of manufactured commodities are included in consumers' goods. Finally, market values of miscellaneous (manufactured) goods plus the totals of value added in printing and publishing[19] are recorded as the miscellaneous category of consumers' goods.[20] Total values of electricity and gas are excluded from our classification of finished goods.

The average annual values of consumers' goods and producers' goods are summarized in Tables 26 and 27. The percentage distribution of the two types of goods is presented in Table 28. According to these figures, finished goods available for domestic use increased at a much higher rate than the growth of domestic production. Along with the rapid increase of finished goods available for domestic use, there was a drastic change in their composition between consumers' and producers' goods. The percentage share of producers' goods rose from 3.6 to 18.1 percent during 1919/21–1939/40. Among the manufactured goods available for domestic use, the relative share of consumers' goods declined from 84.6 percent in 1919–1921, to 66.6 percent in 1939–1940.

Level and Structure of Consumers' Goods

Table 29 presents the per capita level of consumers' goods by type of goods. According to these figures, the level of per capita consumption as a whole increased by 35.9 percent during 1919/21–1939/40. While the level of food available for domestic use shows only a slight increase, the remaining consumers' goods indicate a substantial rise during the period in review.

The composition of consumers' goods also underwent drastic changes, as shown in Table 30. First, the relative share of manufactured goods in total consumers' goods increased from 20.5 to 44.0 percent during 1919/21–1939/40. Second, the share of manufactured goods in total foodstuffs rose from 8.5 to 24.9 percent during the same period. Third, the ratio of clothing to foodstuffs showed considerable improvement, as would normally be expected during a period of higher consumption.

There are at least two basic questions raised by these findings. First, to what extent do the findings represent the true magnitude

Table 26

FLOW OF CONSUMERS' GOODS TO DOMESTIC USE
BY INDUSTRIAL ORIGIN, 1910–1940
(million K¥ at current prices)

	Annual Averages				
	1919–1921	1924–1926	1929–1931	1934–1936	1939–1940
Agriculture					
Domestic Production	939.3	869.1	519.0	732.1	1,151.9
Exports	116.6	204.6	155.2	267.8	140.6
Imports	20.3	57.9	27.1	54.2	79.6
Flow to Domestic Use	843.0	722.4	390.9	518.5	1,090.9
Forestry					
Domestic Production	35.3	56.3	56.2	74.4	121.2
Exports	-	-	-	-	-
Imports	-	-	-	-	-
Flow to Domestic Use	35.3	56.3	56.2	74.4	121.2
Fishery					
Domestic Production	52.0	65.9	68.7	93.5	236.4
Exports	17.4	24.2	22.1	29.0	76.7
Imports	2.9	5.4	4.2	7.1	11.2
Flow to Domestic Use	37.5	47.1	50.8	71.6	170.9
Manufacturing					
Food					
Domestic Production	71.1	93.8	92.9	169.0	350.9

Exports	1.8	6.1	2.5	6.8	8.4
Imports	12.7	23.6	27.6	35.7	71.6
Flow to Domestic Use	82.0	111.3	118.0	197.9	414.1
Textile Products					
Domestic Production	62.5	61.3	40.7	67.0	168.0
Exports	6.4	22.7	26.9	22.6	40.4
Imports	59.7	103.6	83.8	103.4	194.6
Flow to Domestic Use	115.8	142.2	97.6	147.8	322.2
Others					
Domestic Production	30.1	70.2	82.5	127.2	282.3
Exports	3.0	4.9	9.7	19.5	51.2
Imports	11.9	19.3	23.3	33.2	118.8
Flow to Domestic Use	39.0	84.6	96.1	140.9	349.9
Total Flow of Manuf. Goods to Domestic Use	236.8	338.1	311.7	486.6	1,086.2
Total Flow of Consumer Goods	1,152.6	1,163.9	809.6	1,151.1	2,469.2

Sources: Tables B-1, B-2, B-4, B-5.

Table 27

FLOW OF MANUFACTURED PRODUCERS' GOODS TO DOMESTIC USE, 1919-1940
(million K¥ at current prices)

	Annual Averages				
	1919-1921	1924-1926	1929-1931	1934-1936	1939-1940
Metal Products					
Domestic Production	13.0	14.1	6.7	10.9	27.5
Exports	-	-	-	-	-
Imports	8.7	13.7	25.6	39.4	137.4
Flow to Domestic Use	21.7	27.8	32.3	50.3	164.9
Machine & Tools					
Domestic Production	5.7	6.3	9.7	11.5	64.9
Exports	-	1.1	1.8	2.8	1.0
Imports	13.4	17.2	26.9	56.0	183.6
Flow to Domestic Use	19.1	22.4	34.8	64.7	247.5
Chemical Products					
Domestic Production	0.2	5.9	13.1	22.9	100.1
Exports	-	-	-	-	-
Imports	0.1	-	0.8	0.9	2.0
Flow to Domestic Use	0.3	5.9	13.9	23.8	102.1

Ceramics (Cement, Lime, etc.)					
Domestic Production	0.2	0.6	6.2	9.2	29.3
Exports	0.3	0.5	1.7	3.8	2.6
Imports	2.1	2.4	3.1	5.9	4.7
Flow to Domestic Use	2.0	2.5	7.6	11.3	31.4
Total Flow of Producers' Goods	43.1	58.6	88.6	150.1	545.9

Sources: Tables B-3, B-6.

Table 28

PERCENTAGE SHARE OF CONSUMERS' GOODS AND PRODUCERS' GOODS IN DOMESTIC USE
(based on current prices)

	Annual Average		
	1919–1921	1929–1931	1939–1940
Total finished goods	100.0	100.0	100.0
Consumers' goods	96.4	90.1	81.9
Producers' goods	3.6	9.9	18.1
Manufactured finished goods	100.0	100.0	100.0
Consumers' goods	84.6	77.9	66.6
Producers' goods	15.4	22.1	33.4

Sources: Tables 26 and 27.

of per capita consumption for the majority of inhabitants in the country? Second, what are the determining factors that caused the drastic changes in the composition of consumers' goods in favor of manufactured goods?

Let us begin with the first question. Since foodstuffs generally occupy the dominant position in private consumption of low-income countries, an investigation of long-term trends in per capita use of foodstuffs may shed some light on this question. In this regard, one would normally expect a rising trend in per capita use of food grains during a period of income rise, even though the income elasticity of demand for food is usually less than one.[21]

Contrary to the higher level of per capita consumption in Table 29, per capita consumption of food grains declined during the late 1920s and the early 1930s, as shown in Chapter V. It appears that the increase in per capita consumption of manufactured food compensated for the reduction of food grains during the 1930s, but the compensation was not sufficient to produce any significant rise in the per capita consumption of food as a whole. If we estimate per capita income by NVA (net value added), the income elasticity of demand for foodstuffs during 1924/26–1939/40

Table 29
PER CAPITA LEVEL OF CONSUMERS' GOODS
FOR DOMESTIC USE
(K¥ in 1936 prices)

Annual Averages	Foodstuffs[a]	Clothing[b]	Others[c]	Total
1919–1921	33.7	5.0	3.1	41.8
1924–1926	30.7	6.5	6.1	43.3
1929–1931	29.0	5.1	8.0	42.1
1934–1936	29.4	8.4	11.6	49.4
1939–1940	35.7	8.7	12.4	56.8
Wider Intervals				
1919/21–1924/26	32.2	5.7	4.7	42.6
1924/26–1929/31	29.9	5.8	7.1	42.8
1929/31–1934/36	29.3	6.8	9.8	45.9
1934/36–1939/40	32.5	8.6	12.0	53.1

Sources: Tables B-1, B-2, B-4, B-5.

a. Includes agriculture, fishery, and manufactured foods.
b. Includes all the textile products (manufactured).
c. Includes forestry products and manufactured consumers' goods except foods and textiles.

appears to be roughly 0.11. This unusually low value will be explained fully in the next chapter.

The contrasting trends between per capita use of manufactured and non-manufactured goods might have been caused, at least in part, by the limitations of the data used here. In measuring long-term trends in private consumption, our estimates for manufactured consumers' goods may involve far more upward bias than those for agricultural products. This is because there was a substantial expansion over time in the scope of measurement. As mentioned in Chapter I, during the early years of the colonial period, most goods consumed within a household were produced by the household itself.[22] The later years of the colonial period, however, witnessed some basic changes in this regard. Rapid development of transportation facilities and communications, along with the creation of various economic institutions,[23] made it easy to distribute

Table 30

PERCENTAGE COMPOSITION OF CONSUMERS' GOODS

1919-1940

(based on producers' current prices)

	Annual Averages				
	1919-1921	1924-1926	1929-1931	1934-1936	1939-1940
Manufactured Goods vs. Non-Manufactured Goods					
Manufactured Goods	20.5	29.0	38.7	42.3	44.0
Non-Manufactured Goods	79.5	71.0	61.3	57.7	56.0
Total	100.0	100.0	100.0	100.0	100.0
By Type of Goods					
Food Stuffs	83.5	75.7	69.0	68.4	67.9
Crude[a]	91.5	87.4	78.9	74.9	75.3
Manufactured	8.5	12.6	21.1	25.1	24.7
Clothing	10.0	12.2	12.0	12.8	13.1
Others	6.5	12.1	19.0	18.8	19.0
Total	100.0	100.0	100.0	100.0	100.0

Source: Computed from Table 26.

a. Includes agriculture and fishery.

manufactured goods, including imports, throughout the country. The government policy of developing Korea as a major market for Japanese manufactured goods facilitated these changes. At the same time, the increasing intensity of farming, coupled with the exodus of workers from rural areas, forced farmers to abandon many of their subsidiary activities previously carried out to meet the need for non-agricultural products. Subsidiary activities in off-season periods also tended to be concentrated on the production of a few items of high demand, such as cocoon products, straw rope, and bags; this was partly directed by public policy. To the extent that consumers' goods produced by subsidiary activities of households were replaced by marketed goods (mostly manufactured goods)[24], there is an upward bias during the later years of the colonial period because most of the subsidiary activities were not recorded in official data. While we cannot estimate the extent of the upward bias, the effects of the extended coverage must be taken into consideration when the figures of Tables 29 and 30 are evaluated.

Another source of this upward bias may be the rapid urbanization of the 1930s. It is well known that the scope of subsidiary activities within a household is far more limited in urban areas than in rural areas. In addition, the cost required to maintain a certain standard of living tends to be higher in urban than in rural areas.[25] Thus, the rapid urbanization itself must have given some degree of upward bias in our estimates of manufactured consumers' goods, when they are used to measure the true level of per capita consumption of manufactured goods during the colonial period.

In addition to the possible upward bias in our estimates, the implications of inventories should also be examined, because our estimates of manufactured consumers' goods include inventories. Earlier, we pointed out Japan's active preparation for war, beginning from around the middle 1930s, which resulted in an increase in commodity reserves for wartime needs. Even though we cannot measure the exact magnitude of inventory build-up connected with war preparation, it seems reasonable to assume a rapid increase of inventories during the later years of the colonial period.

To sum up the level and structure of consumers' goods consumption, the relatively constant level of per capita food available for domestic use during the period in review seems to indicate a contrasting pattern between domestic production and private consumption. Rapid growth of domestic production was not accompanied by a corresponding improvement in private consumption for the majority of inhabitants, in spite of the low level of per capita income prevailing during the early years of the colonial period.

Level and Financing of Producers' Goods

If capital formation is defined "as the flow of currently produced commodities and services into the stock of economic goods",[26] then it is composed of producers' goods, inventories in producers' hands, and construction. Thus the scope of producers' goods (including building materials) in the present analysis, covers only a portion of capital formation. However, in the absence of relevant data for the estimates of capital formation during the colonial period, the analysis of producers' goods will reveal some characteristics relating to capital formation.

Table 28 indicated that the share of producers' goods in total finished goods increased substantially during 1919/21–1939/40. This rising trend indicates that the rapid growth of the commodity-product plus the net surplus of international commodity flow moved progressively in favor of capital formation. The share of producers' goods in net commodity-product was increased from 3 percent to over 14 percent, suggesting a substantial rise in domestic capital formation during this period. Thus it may be argued that, in relative terms, the large increase in finished goods available for domestic use during the later years of the colonial period contributed far more to capital formation than to domestic consumption.

Let us consider next the financing of producers' goods in terms of external and internal sources. As for external sources, real capital inflow from Japan was estimated at roughly 18 million K¥ during the late 1910s and 142 million K¥ during the late 1930s,[27] whereas the total values of producers' goods were 43 million K¥ and 546 million K¥ respectively in Table 27.[28] Thus, the ratio of real capital inflow to producers' goods was reduced from 58 to about 20 percent during this period. This seems to indicate that domestic savings accounted for an increasing portion of producers' goods during the later years of the colonial period, particularly during the industrialization of the 1930s.

Domestic sources of finance for producers' goods may be divided between private savings and government financing. Since the private sector played the major role in the growth of commodity-production during the colonial period, government financing is excluded from the present analysis. The long-term trends in private savings may be inferred by observing the ratio of deposits at financial institutions to total commodity-product, as shown in Table 31. According to these figures, the savings ratio shows a substantial increase during the entire period in review, particularly during the late 1930s, implying an even higher marginal propensity to save. The major reason for the higher level of private savings should be attributed largely to the new pattern of income distribution during the later years of colonial period, as discussed in Chapter IX.

Table 31

TOTAL DEPOSITS, TOTAL COMMODITY-PRODUCT, AND SAVINGS RATIOS
1920–1939
(million K¥ in current prices)

Annual Averages	Annual change in Deposits[a]	Commodity-Product	Ratios, (2) to (3) in Percentages
(1)	(2)	(3)	(4)
1925–1929	109	1,210	9.0
1930–1934	149	1,002	14.8
1935–1939	381	1,812	21.0

Sources by Column:

(2) Government-General of Chōsen, *Chōsen kinyū jijō sankosho.*
(3) Tables A-3, A-7, A-8, A-9.

a. Total deposits at banks, financial associations, and postal deposits.

Chapter V

AGRICULTURAL SECTOR ANALYSIS

The magnitude and characteristics of Korean economic growth during the colonial period will be further illuminated if the analysis of its over-all growth is now supplemented by sector analysis. The analysis will concentrate on the agriculture and manufacturing sectors, because their combined products account for over 80 percent of total commodity-production.

Growth and Composition of Output

Let us first consider the composition of agricultural product. Table 32 presents the percentage distribution of the agricultural product according to type of output. Even though the figures for the relative share of crops vary over the period under review, it is evident that Korean agriculture was largely concerned with crop production.

Among the various crops, the production of rice dominated; its percentage share among the five major crops rose from 43.8 to 55 percent during 1910–1939. This trend may be explained in terms of demand structure. Rice was traditionally preferred to all other grains in Korea. Thus its production accounted for the major portion of total agricultural product even before the colonial period. After its annexation to Japan, the Korean economy was developed into the major source of food for Japan. Given the relatively similar preference for rice over other grains in the diets of the two countries, it is not surprising to find that the dominant position of rice production was enhanced, and rice became Korea's major export product.

The growth of the agricultural product as a whole may be measured either by gross product or by net value added (NVA); the former includes all costs of production, whereas the latter excludes them. However, since the relative share of cost in total value changed a great deal over the period, as shown in Chapter II, NVA

Table 32

PERCENTAGE DISTRIBUTION OF AGRICULTURAL PRODUCT
(based on market values, current prices)

Annual Average	Crops	By-Products	Live-stock	Manure	Cocoon Products	Total
1910–1914	90.1	0.4	6.3	2.8	0.4	100
1915–1919	87.2	2.8	6.3	2.9	0.8	100
1920–1924	88.3	3.0	5.4	2.8	0.5	100
1925–1929	82.3	3.6	4.9	7.5	1.7	100
1930–1934	79.3	4.5	3.8	10.8	1.6	100
1935–1939	81.7	3.6	4.8	8.1	1.8	100
1940	81.1	3.9	6.4	6.3	2.3	100

Source: *Chōsen tōkei nempō.*

must be used for the analysis of growth trends.

In view of the dominant position of rice production in Korean agriculture, the analysis of relevant data for rice production may reveal the major sources of growth for the agricultural product as a whole. Table 33 is prepared for this purpose. If we define the traditional inputs of agriculture as land and labor, it is clearly shown in the table that the growth of rice production during the early years of the colonial period was attributable both to increase in the quantity and productivity of traditional inputs. However, the growth after the mid-1920s was due more to increased productivity than to quantity change as the Korean agriculture became more intensive.

This pattern of growth may be explained in terms of the major types of improvements made in the agricultural sector during the colonial period. In this connection, it is helpful to observe the types of improvement made in Japanese agriculture, because the agricultural techniques of Japan were directly applied to Korean agriculture during this period. There were two types of improvement in Japanese agriculture:

> The first was land improvement, including better irrigation and drainage facilities and the reclamation of some arable

land—mostly paddy rice fields. The second encompassed superior seeds, better methods of crops cultivation, and increased input of manures and fertilizers.[1]

Table 33
PRODUCTIVITY INDICES OF LAND AND LABOR IN RICE PRODUCTION (1910-1940)

Year	Volume of Rice Production	Paddy Field	Labor in Agriculture (male equivalent)	Land Productivity (1)/(2)	Labor Productivity (1)/(3)
	(1)	(2)	(3)	(4)	(5)
1910-1914	87	93	-	94	-
1915-1919	100	100	100	100	100
1920-1924	108	102	101	106	107
1925-1929	119	104	106	115	112
1930-1934	148	110	101	135	147
1935-1940	165	107	97	154	168

Sources by column:

(1) Taken from the statistical yearbooks of the Government-General, except 1924-1935 data from Table 1.
(2) Taken from the statistical yearbooks of the Government-General.
(3) Table 34.

During the early years of the colonial period, the major emphasis in developing Korean agriculture was placed on the first type of improvement under the direction of the Government-General and of semi-official institutions such as the Oriental Development Company. To these considerations should be added the fact that the so-called "slack" in the rural economy at the time of Korea's annexation to Japan existed mainly in the form of wasteland that could easily be reclaimed. The Government-General also established some new institutions designed to develop agriculture, such as agricultural schools and model farms, but their effects were not yet widespread. In short, the first stage of development in the Korean agriculture was to promote extensive farming.

As for the second type of improvement mentioned in the above quotation, a nationwide technological diffusion in the rural areas took place only after 1920, coinciding with the launching of an extensive plan by the Government-General for increasing production. In this regard, it may be said that the agricultural transformation into intensive farming started from the middle 1920s.

Agricultural Transformation since 1920

The Policy of the Government-General[2]

When the thirty-year plan for increasing rice production in Korea was launched in 1920, the original goal was to expand and improve 800,000 *chungbo* out of total paddy land of 1,548,000 *chungbo* in 1920. During the first half of the plan period, the goal was to obtain a net increase of 9,200,000 *suk* in rice production (the rice production of 1920 was 14,882,000 *suk*). In pursuit of these targets, the role of the Government-General included: 1) attaining necessary information, i.e., surveys for potential land improvements, irrigation facilities, research, and so forth; 2) providing for subsidies to landowners for improvements; 3) establishing special agencies to assist and supervise projects; and 4) encouraging the diffusion of improved seeds. While the Government-General assumed the responsibility for performing these tasks, the landlord class was expected to undertake the various projects.

However, the implementation of the plan soon encountered a serious difficulty arising from the landlords' lack of incentives for reclamation. After an initial expansion of the cultivated area (utilizing the easily reclaimable land) during the early years of development, further expansion confronted higher unit costs. The 1920s plan failed to offer incentives conducive to continuing reclamation.

Faced with this difficulty, the Government-General revised its original plan and announced a ten-year plan in 1926. The main features of the revised plan were: 1) providing low-interest loans for land reclamation and improvements; and, 2) placing the major emphasis on the diffusion of technological improvements in cultivation (that is, extensive use of new inputs such as fertilizers, improved seeds, and so forth.) Thus the revised-plan period marks the beginning of an extensive diffusion of technology in Korean agriculture. The ten-year plan was discontinued in 1934, in the face of severe agricultural depression in Japan.

Productivity Changes

The implementation of these plans for increasing rice production brought fundamental changes in the traditional structure of Korean agriculture, both in inputs and institutions.

Table 34 summarizes the long-term trend in the use of the major inputs of agricultural production in Korea. Before the analysis, however, it may be desirable to explain briefly the nature of the figures in Table 34. The land input covers both paddy field and dry land. In order to account for the productivity differences between them, they were weighted according to the average price difference between paddy field and dry land during 1930–1935.

As for the labor input, the only data available for the entire period under review are the Government-General's year-end estimates of population distribution by occupation. These are classified according to sex and according to the following categories: 1) gainful workers who are solely engaged in agriculture; 2) gainful workers who also have other occupations; 3) the unemployed. Two assumptions are made for the derivation of labor-input index from the official data. First, we assume that a female worker is equivalent to one-half a male worker, measured by total production per year, because the average female works fewer hours than the average male throughout the year. Second, it is also assumed that "gainful workers with other occupations" devoted only one-fourth of their working hours to agricultural production. In this way, an index of labor input is constructed (Table 34), which will at least indicate the broad trend of labor inputs into agricultural production.

The fertilizer input includes both manure and commercial fertilizers. These are weighted according to the relative price differences for deriving the fertilizer index. The traditional input index refers to the index of land and labor plus manure. Finally, the total input index is the aggregate of all inputs using the geometric formula of a Cobb Douglas production function. The indices for each input are weighted by the respective shares in the total cost of production. For this purpose, the cost data of rice production in 1933 are used.

Apropos of Table 34, it is quite evident that agricultural transformation after 1920 was characterized by the rapid increase of new inputs such as fertilizer and by the diffusion of agricultural technology as measured by the spread of improved seeds. From the middle 1920s, the rise of the total input index was largely brought about by the tremendous expansion of a new input (fertilizer), while the traditional inputs showed only a moderate increase or even a declining trend. Incidentally, the construction of irrigation facilities was not one of the major factors contributing to the agricultural transformation. The percentage of cultivated areas for rice production equipped with irrigation facilities (constructed) was only 4.5 percent in 1926, 8.6 percent in 1930, and 12.7 percent in 1936.

In order to examine the productivity change of inputs, Table 35 is presented. The figures indicate once again that the rapid increase in traditional input productivity occurred from the 1930s and it should be attributed to the extensive use of new inputs. At the same time, this table also reveals that the average productivity of new inputs decreased from the middle 1920s. The explanation for this must be deferred until we examine the institutional changes that took place in the rural economy during the colonial period.

Major Institutional Changes

The transformation of Korean agriculture in accordance with the colonial policy of the Government-General required some basic changes in the institutional setting. The major institutional changes may be examined in connection with the Japanese immigrants to Korean farms and with the landlord-tenant relationship.

Even before the formal annexation of Korea in 1910, the

Table 34

QUANTITY INDICES OF AGRICULTURAL INPUTS

Year	Land Input (1)	Labor Input (2)	Fertilizer Input (3)	Traditional Input (4)	Total Input (5)	Percentage of Paddy Land Using Improved Seeds (6)
1915-1919	100	100	100	100	100	39
1920-1924	112	101	168	107	107	64
1925-1929	114	106	457	110	133	73
1930-1934	117	101	736	109	144	77
1935-1940[a]	119	97	1,129	108	156	85

Notes by column:

(1) Paddy field and dry land, weighted by average price differences during 1930-1935.

(2) Male worker-equivalent of total gainful workers with the following assumptions:
   a. a female worker is equivalent to one-half of a male worker.
   b. a part-time worker in agriculture is equivalent to one-fourth of a full-time worker.

(3) Manure and commercial fertilizers, weighted by the relative prices.

(4) Includes land, labor, and manure, computed of the same method as (5).

(5) Aggregate index of all inputs.
Aggregate indices are computed by the geometric formula of a Cobb-Douglas production function, with each factor weighted by its share in the 1933 rice production cost.

a. (excluding 1939)

Japanese government encouraged farmers to emigrate to Korea and attempted to place the Japanese in a strategic position to control Korean agriculture. There were two major channels through which the Japanese immigrants obtained arable land in Korea: the arrangements of semi-official institutions, and the purchase of land by individuals. The Japanese immigrants of the former type

Table 35

INDICES OF OUTPUT AND FACTOR PRODUCTIVITY

Periods	Output[a]						Factor Productivity[d]	
	Rice	Barley	Beans	Other Grains[b]	Cotton	Total Crops[c]	Traditional Inputs	Total Inputs
1915–1919	100	100	100	100	100	100	100	100
1920–1924	108	101	105	113	173	110	103	103
1925–1929	119	101	101	103	229	114	104	86
1930–1934	148	112	101	98	220	129	118	90
1935–1940	165	132	102	96	288	145	134	93

a. Each crop index is computed from *Chōsen keizai nempō, 1939,* Appendix, pp.6–7.

b. Includes such crops as wheat, corn, peas, potatoes, etc.

c. The index of total crops is computed using the market prices of each period as relative weights.

d. Index of total crops divided by input index of Table 34.

were called "protected settlers," and the latter "free settlers."

Among the new institutions designed to facilitate Japanese immigration to Korean farms, the most important one was Toyo takushaku keisha (the Oriental Development Company). It was established as a joint-stock company of Koreans and Japanese, but its management was completely Japanese. According to the policy guidelines of the company, Korean agriculture was to be developed through the settlement of Japanese farmers in Korea.[3] The company undertook the task of bringing to Korea over 1,000 Japanese farm-households annually beginning in 1911. The company purchased arable land from Koreans and distributed to the Japanese settlers two *chungbo* per household until 1921. The allotment of arable land was increased to over five *chungbo* in 1922, and most of it was leased to Korean tenant-farmers.

As for the so-called "free-settlers," most of them acquired arable land through wealth accumulated from the loan business or through trade within Korea rather than from capital inflow from Japan. During the transitional period of 1876–1910, Korean exports consisted exclusively of agricultural products, and foreign trade was handled largely by Japanese merchants in Korea. They often took advantage of social disturbances and the increasing hardship of farm-households by purchasing agricultural crops prior to the harvest time at bargain prices or by lending money to farmers at high interest rates with arable land as the collateral. Thus, most Japanese landlords of the "free-settlers" category in Korea were formerly small merchants or usurers.

There were also a few Japanese landlords of an entrepreneur type who from the outset made lump-sum investments in landownership to obtain the high return realizable from rice production. With the land-tenure system so favorable to landlords in Korea, ownership of land was the most attractive investment project until the industrialization of the 1930s. At the same time, there was a rising tendency among a few landlords to consolidate small holdings. The concentration was largely in the hands of Japanese landlords, because of their favorable social status in Korea and because of their easy access to the funds necessary for consolidation.[4]

In short, the transition of Korean agriculture into a highly productive sector was accompanied by an increase in the number of Japanese landlords, their concentration of landownership, and the presence of entrepreneur-type landlords for the export market. The major implications of these changes in Korean agriculture are well summarized in the following:

> A large portion of the rice land in Korea was owned by large-scale commercial owners, principally Japanese, who collected heavy land rents payable in kind. Most of the rice collected as rent moved into the export trade, and about 60 percent of Korea's exports came from these large-scale farmers.[5]

The status of large-scale Japanese landlords in Korean agriculture is shown in Table 36.

Let us turn next to the land-tenure system of the colonial period. By the time of annexation in 1910, Japan had successfully transformed her traditional agriculture into a highly productive

Table 36

TOTAL NUMBER OF LANDLORDS BY AMOUNT OF LAND AND NATIONALITY, 1921–1935

	Over 200 *chungbo*		200–100 *chungbo*	
Year	Koreans	Japanese	Koreans	Japanese
1921	65	169	360	321
1925	45	170	344	360
1930	50	187	304	361
1935	45	192	315	363

Source: Computed from the official data on land-tax payers by size of land by nationality. Kobayakawa, *Chōsen nōgyō hattatsu-shi,* Table A-4.

sector. Thus the Government-General in Korea was well equipped with the techniques and policy instruments conducive to agricultural development. Instead of reproducing Japan's approaches to agricultural development in Korea, however, some modifications were required by the specific objectives of developing Korean agriculture. These modifications are clearly reflected in the land-tenure system of the colonial period.

As already noted, the main objective of developing Korean agriculture was to provide a food supply for Japan. For this purpose, it was essential to transform the Korean agriculture into a highly productive sector and, at the same time, to obtain complete control of the distribution of output so as to maximize the surplus output that was to be channeled into Japan. To these requirements should be added the over-all policy objective of the Government-General, which was to discourage the development of non-agricultural sectors until the 1930s. Thus, the net increase in the labor force during the period under review had to be absorbed largely by the agricultural sector.

Under these circumstances, it is not surprising to find that the approach taken by the Government-General with respect to the development of Korean agriculture was to develop it from "above," through the landlords' pressure on, and control of, the peasants. It was simply not possible to achieve these objectives by developing Korean agriculture through any schemes that might have required initiative on the part of the peasants. With these objectives, the land survey of 1912–1918 was carried out, and a new land-tenure system established.[6]

In essence, the new system legalized the formation of a landlord class, and increased this landlord class's share of total arable land.[7] The latter trend continued throughout the entire period, as shown in Tables 37 and 38. We mentioned earlier that the deterioration of the rural economy during the later Yi dynasty was attributable to the institutional setting that allowed the imposition of undue burdens on the peasants. The land-tenure system of the colonial period did not change the old system, but had the effect of legalizing it.

Table 37

PERCENTAGE DISTRIBUTION OF FARM HOUSEHOLDS BY TYPE OF OWNERSHIP 1913-1939

Annual Average	Proprietors	Proprietors and Tenants	Pure Tenants
1913-1917	21.7	38.9	39.4
1918-1922	20.4	39.0	40.6
1923-1927	20.2	35.1	44.7
1928-1932	18.4	31.4	50.2
1933-1937	19.2	25.6	55.2
1939	19.0	25.3	55.7

Source: Takeo Suzuki, *Chōsen no keizai* (Tokyo, 1942), p. 246

Under the land-tenure system of the colonial period, the precise relationship between a landlord and tenant was determined by an annual contract, with the terms decided according to the bargaining positions of the two parties. The rapid growth of population relative to both the limited expansion of cultivated areas and employment opportunities in non-agricultural sectors intensified competition among the peasants for the tenancy of arable

Table 38

PERCENTAGE DISTRIBUTION OF ARABLE LAND BY TYPE AND OWNERSHIP

Years	Total Paddy Field (1)	Paddy Field by Pure Tenants (2)	(2)/(1) in percent (3)
1914–1917	1,261	829	65.7
1918–1922	1,544	993	64.3
1923–1927	1,566	1,009	64.4
1928–1932	1,620	1,077	66.5
1933–1935	1,671	1,132	67.7

Source: Takeo Suzuki, *Chōsen no keizai* (Tokyo, 1942), p. 246.

land. The land rent was determined by various methods, depending upon the region and the quality of the cultivated area. The most common methods were *tochi* and *tachak.* The former is a variable rent determined annually on the basis of the total crops at harvest time, whereas the latter refers to a rent equivalent to one half of the total amount harvested. Both types of land rent are essentially sharecropping, which may fail to provide farmers with incentives for a higher level of production compared with a fixed-rent system. In view of the weak bargaining position of tenants in the rural economy during this period, it was not surprising to find that the specific rates of land rents were determined in order to maximize the relative share of the landlords.[8] According to an official investigation:

> The maximum amount of rent runs up as high as four-fifths and even nine-tenths of the crop and the minimum amount goes down as low as one-third and one-fifth. The most prevalent amount is about one-half of the yield. In certain localities, however, like Namwon country, North Chulla Province, about four-fifths of the yield is paid in rent. In North Choonchung Province, the prevailing amount of rent is about seven-tenths. These high rents result from a gradual raising of the amount by the landlords at the time of new leases.[9]

In addition to the land rent, various burdens were shifted on to tenant farmers as their bargaining position deteriorated. The most common burdens levied on tenant farmers were: full or partial amount of taxes and fees for irrigation facilities; full or partial costs for seeds, commercial fertilizers, land improvements or repairs; and various service charges. Of special importance here is that the cost of new inputs (that is, improved seeds, commercial fertilizers) was borne by the tenant farmers, while the land rent was determined by the total volume of production.[10] Consequently, tenant farmers were often forced to increase the use of new inputs beyond the "break-even" point as long as it resulted in a net increase in total output. This explains the decreased productivity of new inputs, as shown in Table 36.

The major implications of the land-tenure arrangements for the structure of the rural economy may now be observed. First, the agricultural transformation imposed the heaviest burden on tenant farmers. It was estimated that over 62 percent of total tenant-farm households were short of even the minimum amount of food required for an adequate diet, and the average amount of farm-household debt amounted to 137 *yen* (about 30 percent of annual income) in 1930.[11] This deterioration of rural conditions took place while the over-all growth of the agricultural product was substantial.

Second, landownership was the most profitable investment project in Korea. In addition to the extremely high level of land rent and the shift of incidental burdens to tenants, the landlord class profited from the land-improvement projects in connection with the official plan for increasing rice production. The land improvements yielded capital gains arising from the rapid increase of land value, and reduced the real burden of the land tax. The land-tax system during the colonial period is summarized in the following:

> On the completion of the land survey (in 1918) the Government placed a fixed value on each lot according to its valuation. The standard of valuation was, of course, the productivity

> of the individual lot and also took into account other conditions, local and general, which affect the value of the land. The tax rate, then, was fixed at 1.3 percent of the land value. However, the Government's fixed value of land was set for all time. Accordingly, it takes no account of changes in productivity or fluctuations of the selling value from other causes.[12]

Of special importance is the tax burden based on the fixed value of land, and no land reassessments took place during the colonial period. Accordingly, the official value of land for tax purposes became only a fraction of its market value after land-improvement projects were completed. The initial tax rate of 1.3 percent was also considerably lower than the land tax that the Japanese government levied during the early years of modern economic growth in Japan.[13] Indeed, it was an integral part of the over-all policy of the Government-General to make landownership most attractive. Accordingly, the rate of return from investment in landownership was high, as shown in Table 39.

Major Contributions of Agricultural Development

The main function of Korean agriculture during the colonial period may be examined according to the types of agricultural contributions. In this regard, Professor Kuznets's analysis is very useful;[14] "If agriculture itself grows, it makes a product contribution; if it trades with others, it renders a market contribution; if it transfers resources to other sectors, these resources being productive factors, it makes a factor contribution."

The Product Contribution

We observed in the preceding chapter that the growth in agricultural products accounted for the major portion of the total commodity-product increase—as reflected also in its high contribution rate—until the rapid industrialization of the 1930s.

Despite the substantial growth of agricultural product, however, the relevant data indicate that the growth of food grains lagged behind the potential growth of the domestic demand, let

Table 39

RATE OF RETURN TO LANDOWNERSHIP AND COMMON STOCKS

	1931	1937
Korea		
Paddy field	7.7%	8.0%
Dry land	8.3%	8.5%
Major Common Stocks	6.9%	6.5%
Japan[a]		
Paddy field	3.89%	4.89%
Dry land	3.69%	5.46%

Source: Computed from surveys taken by Chōsen Shokusan Ginkō (Industrial Development Bank of Korea).

a. as of 1936.

alone the foreign demand. For instance, the average annual growth rates of population and per capita commodity-product during the period 1910–1940 were 1.6 percent and 1.5 percent respectively. Thus, if we assume an income elasticity for food of 0.7 percent,[15] the potential demand for food grains would have increased at an annual rate of 2.2 percent, whereas the volume of crops during the same period increased at an annual rate of 1.9 percent.

Let us consider in some detail whether the product growth within the agricultural sector contributed to the welfare of the inhabitants. This may be done by analyzing the effects of the product growth on the domestic food supply. Table 40 shows the per capita use of the major food grains during 1912–1940. The computation of these figures requires explanation. First, we used our estimates of total population and the volume of rice production for the reasons already indicated. Second, it was observed that about 7 percent of the total rice production was used for seeds and brewing during the period 1930–1936. To account for other uses and the loss of rice (for example feeds, industrial uses, waste, extraction losses, and so forth), we assumed that 10 percent of the rice

Table 40

PER CAPITA CONSUMPTION OF MAJOR FOOD GRAINS
(in *suk*)

Annual Average	Rice	Barley	Millet
1912–1915	.64	.37	.24
1916–1920	.61	.40	.28
1921–1925	.53	.36	.32
1926–1930	.47	.34	.33
1931–1935	.46	.36	.26
1936–1940	.54	.41	.24

Sources: Population data, from Table 13.
Rice production, from Table 1 and statistical yearbooks of the Government-General.
All other data, from the statistical yearbooks of the Government-General.

Note: Consumption of food grains is estimated as follows: domestic production + import – export – 10% of domestic production (for other uses).

production was not used for food grain. The same amount, 10 percent, was deducted from barley and millet production to account for uses other than food. Third, the availability for domestic food consumption was finally obtained by further adding the net import surplus to (or subtracting the net export surplus from) the above estimates.

The calorie equivalent of the daily food availability per capita is presented in Table 41.[16] The total daily calories per capita in this table are calculated under the assumption that the calories from the three major crops make up 70 percent of the total food consumed, regardless of changes in per capita income level.[17]

According to the figures in Table 40, the per capita use of food grains as a whole declined substantially after the early years of the colonial period. In particular, the per capita availability of rice declined drastically with the beginning of the agricultural transformation in the 1920s. The reduction in per capita use of rice was not fully compensated for by an increased use of other

Table 41

DAILY PER CAPITA FOOD AVAILABILITY IN CALORIES

Annual Average	Calories from Rice	Calories from Barley	Calories from Millet	Total Daily Calories Per Capita[a]
1912-1915	877	490	274	2,133
1916-1920	842	526	329	2,206
1921-1925	731	468	365	2,033
1926-1930	650	449	381	1,924
1931-1935	630	468	296	1,812
1936-1940	741	549	274	2,033

Sources: Table 40. To convert food from physical quantities to calories, the conversion ratios suggested by the United States Department of Agriculture, *Composition of Food Used in Far Eastern Countries,* Agricultural Handbook No. 34 (1952) are used.

a. It is assumed that the calories from the three major crops make up 70 percent of the total food consumed.

grains, contributing to the reduction in total daily calories per capita as shown in Table 41. Calorie intake reached its lowest ebb during the first half of the 1930s, and then there was a slight upward trend.

Let us attempt to explain the trend revealed in Tables 40 and 41. At the risk of repetition, it should be pointed out once again that the rapid growth in population, without a corresponding expansion in cultivated areas and employment opportunities in non-agricultural sectors, tended to increase competition for tenancy of land among the peasants. The situation progressively increased the burden of tenants during the later years of the colonial period. It was precisely this increasing burden on tenants and small farmers that brought a reduction in per capita consumption of food grains for about 80 percent of farm-households at the time of a rapid growth in agricultural production. This was clearly shown in our analysis of the land-tenure system during the colonial period.[18] To the high rents should be added declining agricultural prices and reduced real wages of laborers during the depression of the 1930s, to explain the ebb in per capita food consumption.

An average tenant farmer in Korea could not secure a year's requirement of food grains from his cultivated area because the size of his allotment was too small and the land rent too high. In these circumstances, the common practice was to sell his share of rice production for larger quantities of inferior grains. In B. F. Johnston's words:

> The heavy burden of rent in kind tended to increase the cash stringency of the small farmers, which in turn had the effect of increasing the percentage of the crop which was commercialized. The magnitude of Korea's exports (of rice) after 1930 is to be explained to a large extent by the low level of living prevailing among Korean farmers and the pressure to sell rice for cash at the expense of their own level of food consumption.[19]

Even the supply of inferior grains was very often in acute shortage around the end of a harvest season (prior to the spring harvest of barley). It is estimated that approximately forty-three kinds of wild grass were used by the Korean farmers to supplement their shortage of food.[20] B. F. Johnston points out that "in a real but complex sense, Korea's exports during the 1920s and early 1930s were 'forced' exports. The phrase 'starvation exports' was quite commonly applied to Korea's exports by Japanese food officials and agricultural economists."[21]

In short, the evidence leads us to conclude that the product growth in connection with the agricultural transformation of Korea had an adverse effect on the availability of domestic food until the middle 1930s. If the poverty of farmers in Japan and Korea in the early 1930s is compared, the fundamental differences in the nature of their poverty may be pointed out: the poverty of Japanese farmers meant a reduction in money income to the extent of suffering a deficit in the family budget, whereas the poverty of Korean farmers involved the prospect of starvation and a threat to survival.

This observation is the basis of our assumption in Chapter II that the magnitude of underreporting in the official data of rice production is positively correlated with the declining trend in the availability of per capita food grains.[22]

The upward trend in per capita use of food grains during 1936–1940 should be interpreted with due qualification. To the extent that the rapid industrialization of the 1930s rendered fundamental changes in the industrial structure, particularly in the industrial distribution of the labor force which was shifting to non-agricultural sectors, per capita consumption of food grains for the country as a whole must have increased. However, as shown in Chapter IV, the industrialization of the 1930s failed to produce any drastic changes in the industrial distribution of Korean workers.

The upward trend may be partly attributable to the new policies of the Government-General designed to check the deteriorating conditions in rural areas. Recognizing a serious threat to the survival of tenant farmers, the government attempted to regulate land-tenure arrangements and to prevent the physical as well as the moral deterioration of farmers. Measures taken included the Regulation of Agricultural Land (1934) and the nationwide movement for "Advancement of Rural Areas," beginning around the middle 1930s.

The measures designed to protect tenant farmers, however, were soon confronted by the strong opposition of the landlord class, and, after the beginning of war with China in 1937, they were replaced by the policy of "total mobilization" of preparations for World War II.

The main reason behind the upward trend of the period 1936–1940 in Tables 40 and 41, may be the effects of inventory increase. Beginning with 1937, the Government-General tightened its control of the Korean economy as a part of the all-out effort of Japan in preparation for World War II. This was done initially by introducing a license system of foreign trade (regulating both exports and imports), by carrying out the rationing of selected commodities, and by price controls. In this way, domestic consumption was reduced to a minimum, with a corresponding increase in the commodity reserves in preparation for war, and food grains were no exception to the general trend toward a "war economy."[23] At the same time, the military venture of Japan into Manchuria and China increased substantially the demand for food grains, and this was met by the reduction of exports to Japan. Unfortunately, we cannot account for the effects of commodity reserves including food grains and of the consumption of the Japanese military force outside Japan in our computation of per capita availability of food grains. Nevertheless, it may well be said that the per capita use of food grains during the period 1936–1940 did not increase substantially.

In short, the transformation of Korean agriculture into a highly productive sector was accompanied by the reduction of per capita food consumption for the majority of inhabitants, the exceptions being minority groups of Japanese residents and Korean landlords. From the standpoint of development economics, the decline in per capita food consumption was a form of "forced savings," attained by the ruthless land-tenure system of the colonial period. Thus the real significance of the product contribution must be examined in connection with the way these "forced savings" were used.

The Market Contribution

Needless to say, the most important role of Korean agriculture during the colonial period is connected with its market contribution. As mentioned, agricultural development in Korea was geared to Japan's need for food grains, and imports of Korean rice to Japan played a key role in ameliorating Japan's food shortages after the "rice riots" of 1918. The relative amount of Korean rice in Japan's total import of rice is shown in Table 42. In this connection, it should be noted that, unlike the imported rice from other countries, the quality of Korean rice was equivalent to Japanese rice, especially after the widespread use of improved seeds, thus constituting a true supplement to the domestic production of rice in Japan. The transformation of Korean agriculture was accompanied by a quality change designed to obtain better marketability for the output.

Table 42
PERCENTAGE DISTRIBUTION OF JAPAN'S IMPORT OF RICE BY NATIONAL ORIGIN

Period	Import from Korea	Import from Taiwan	Import from Foreign Countries
1911-1915	24	20	56
1916-1920	37	18	45
1921-1925	47	17	36
1926-1930	57	21	22
1931-1935	65	30	5
1936-1938	63	35	2

Source: Bruce F. Johnston, *Japanese Food Management in World War II* (Stanford, 1953) p. 51.

The expansion of rice production alone, however, was not sufficient to meet the increasing demand for rice from Japan. This may be shown by the trends in the ratio of exports of rice to its domestic production as presented in Table 43. The figures indicate that rice exports to Japan increased at progressively higher rates than the growth of domestic production except in the late 1930s, when the preparation for war forced a cut in Japan's import of rice. Japan's demand for rice during the colonial period was met by both the expansion of rice production and a substantial reduction in the relative share of output for domestic consumption. The latter result was obtained through the peculiar land-tenure arrangements which "tended to increase the cash stringency of the small farmers, which in turn had the effect of increasing the percentage of the crops which was commercialized."[24]

The Factor Contribution

The colonial aspect of agricultural development in Korea is most apparent in the analysis of factor contributions, which are concerned with the transfer of resources from agriculture to non-agricultural sectors.

Table 43

RICE PRODUCTION AND EXPORTS

Annual Average	Production Volume	Production Index	Exports (volume in thousand *suk*) Volume	Exports Index	(4)/(2) (in percent)
(1)	(2)	(3)	(4)	(5)	
1912–1915	12,488	100.0	1,181	100.0	9.5
1916–1920	14,101	112.9	2,154	182.4	15.3
1921–1925	14,784	118.4	3,942	333.8	26.7
1926–1930	17,234	138.0	6,022	509.9	34.9
1931–1935	20,562	164.7	8,515	721.0	41.4
1936–1940	21,246	170.1	7,041	596.2	33.1

Source: The Bank of Korea, *Annual Economic Review of Korea,* 1948, statistics, p. III-28.

Let us first consider the capital transfer. We've already mentioned that a substantial amount of "forced surplus" was created by the land-tenure arrangements in the course of the agricultural transformation. However, the surplus of agricultural product remained within the agricultural sector instead of being extracted for the financing of industrial development, as Chapter VI will show.

As for the transfer of labor from the agricultural sector, the only data available are summarized in Table 44. It should be noted here that the year-end estimate of rural population in 1910 is the least reliable because of the infancy of the administrative setup of the Government-General. Nevertheless, it may be true that the growth rate of rural population was slighly higher than that of the total population during the early years of the colonial period, reflecting the immediate effects of a colonial policy designed to achieve agricultural development and a higher rate of natural increase in rural than in urban areas.

Of special importance in this table is a growth rate of rural population that is slower than the growth rate of the total population beginning with the late 1920s. The trend seems to indicate

Table 44

TOTAL AND RURAL POPULATION

Year	(1) Total Population Number (1,000's)	(1) Total Population Index	(2) Rural Population Number (1,000's)	(2) Rural Population Index
1910	14,766	86	10,449	71
1920	17,264	100	14,744	100
1930	20,438	118	15,853	108
1940	23,547	136	16,772	114

Sources by column:

(1) Table 13.

(2) Official data, statistical yearbooks of the Government-General.

the transfer of labor to productive uses in the non-agricultural sector. However, as shown in Chapter IV, the rapid industrialization of the 1930s had no significant effect on the pattern of employment of Korean workers. Thus the trend should be attributed to the large-scale exodus of farmers to urban areas as laborers rather than to this industrialization. The real contribution with respect to the labor transfer was made by the large emigration of Korean workers to other parts of the Japanese Empire beginning with the late 1920s, as discussed Chapter III. Most of them left the agricultural sector to avoid the deteriorating living conditions in rural areas at the time of the agricultural transformation.

Chapter VI

MANUFACTURING SECTOR ANALYSIS

The Industrial Growth of 1910–1930

As mentioned in Chapter I, the policy objective of the Government-General relating to the industrial sector during the early years of the colonial period was to restrict the development of domestic industries. The bulk of non-agricultural product for domestic consumption was provided by the subsidiary activities within the household. The demand for highly fabricated products was largely met by imports from Japan.

There was one exception to the above generalization from the beginning of the colonial period—modern factories connected with rice exports from Korea. The process by which the domestic production of rice was channeled into export involved two distinctive steps. First of all, the farmers produced unhulled rice, and it became an export commodity only after it was hulled at the rice mills. Most rice mills were equipped with modern facilities in order to improve the massive handling and channeling of rice to Japan. In contrast to the practice in Japan, where the rice-hulling was largely done by farmers themselves, there developed in Korea a distinctive division of labor between farmers and rice mills. The Korean farmer was simply the producer of raw material for rice mills. Therefore, a large portion of the benefits arising from the rapid expansion of rice exports was shared by merchants and industrialists who owned and operated the mills. Perhaps the origin of a dualistic structure in the manufacturing sector of Korea may be observed here: while the bulk of non-agricultural products was produced by household industries, rice as an export commodity was supplied by the modern facilities of these mills.

During the second decade of the colonial period, economic conditions in Japan led to the relaxation of official control on investments in the non-agricultural sector of Korea, and, at the same time, a complete economic integration between the two nations was achieved, as pointed out in Chapter I. While the direct controls on investment were lifted, however, there were nevertheless some formidable obstacles to developing modern industries during this time.

Let us begin with the conditions of resource flow from agriculture to modern industries. It should be noted from the preceding chapter that a substantial amount of agricultural surplus was created through the agricultural transformation beginning with the middle 1920s. However, the surplus of agricultural product was not channeled into financing modern industries and, as a result, the redundant labor force in agriculture was not transferred to other sectors for productive use. The main reasons for the failure of generating resource flow from agriculture are given below.[1]

Since the basic objective of Japan in the administration of Korea was to develop a periphery specializing in producing food grains, industrial development was to be discouraged. Furthermore, it was the Government-General's approach to the agricultural development of Korea that the landlords instead of the farmers were to play the strategic role and that a large portion of the landlord class in Korea was to consist of Japanese immigrants. In order to implement these objectives, it was essential to make the ownership of land extremely attractive to landlords and to the potential Japanese immigrants to Korea. Accordingly, the land tax and other measures were designed with a view to ensuring a higher return for investment in landownership than in non-agricultural sectors. Thus, there were neither incentives for landlords nor government mechanisms to channel the agricultural surplus to finance capital formation for industrial development. Instead, the surplus was diverted to the consolidation of landownership and to the expansion of Japanese holdings. Herein lie the colonial characteristics of agricultural development.

At the same time, the institutional arrangements necessary for a resource flow from agriculture to industry were not provided by the Government-General. The colonial period witnessed the establishment of modern banks and other financial institutions. Instead of harnessing domestic resources to finance modern industries, however, the major function of the financial institutions was to expand agriculture and trade with Japan. This role was carried out through special banks and semi-official types of financial institutions such as the Bank of Chosun, the Industrial Bank, the Rural Credit Societies, the Oriental Development Company, and so forth. How closely these institutions were following the policy objectives is clearly reflected in Table 45, which summarizes institutional loans according to their uses. Until 1930, only 4–8 percent of the total was made available for industrial use.

The absence of an organized money market in Korea also deterred the transfer of resources from agriculture to industry. Along with the complete economic integration with Japan, the money market in Japan became the major source of investment funds in Korea. However, only a few Japanese corporations could obtain entry through the sale of securities and stocks to Japan's money market. Investment opportunities in securities and stocks for agriculture surplus were not readily available in Korea.

Formidable obstacles to developing modern industries also stemmed from the fact that Korea's complete economic integration with Japan enhanced competition among Japanese industries. The fact that the industrial sector of Japan was "overextended" during the boom period of World War I intensified further the competition with the potential new industries in Korea.

The major characteristics of the manufacturing sector during the period under review may now be summarized. Along with the agricultural transformation in Korea, there emerged certain conditions conducive to the establishment of modern industries such as the agricultural surplus and domestic demand as reflected in the substantial amount of imports from Japan. However, these conditions were outweighed by unfavorable influences such as government policy and institutional conditions opposing industrialization, the lack of incentives for transferring resources, and pressures from the established industries in Japan. In these circumstances, only the active support and protective measures of the Government-General could easily establish modern industries. However, this was certainly contradictory to the overriding policy

Table 45

TOTAL INSTITUTIONAL LOANS BY TYPE OF USE
(thousand K¥)

Year-end	Agricultural Use	Industrial Use	Commercial Use	Miscellaneous Use	Total Loans
1910	743	1,808	18,165	2,237	22,953
	(3)	(8)	(79)	(10)	(100)
1920	22,865	10,420	132,196	16,968	182,449
	(13)	(6)	(72)	(9)	(100)
1930	173,735	16,953	154,704	59,592	404,984
	(43)	(4)	(38)	(15)	(100)
1937	181,321	165,928	319,922	142,401	809,572
	(22)	(20)	(40)	(18)	(100)

Source: Government-General of Chōsen, *Chōsen kin'yū jijō sankoshō*.
Note: Figures in parentheses are the percentage shares.

of the Government-General.

The Beginning of Modern Industries

That the Government-General was against Korean industrialization does not mean that there was a complete absence of modern industries in Korea. As mentioned already, some of the rice mill factories were equipped with modern facilities. To these should be added a few modern industries established in Korea after the end of World War I. How did these industries emerge?

Prior to the industrialization of the 1930s, the total number of industries equipped with modern facilities included six factories in textiles, four in foods, three in ceramics, and one each in papermaking, leather production, and chemicals.[2] In every case, except for one textile factory, they were established by the Japanese with financial support from Japan in order to supplement the existing industries in Japan or to take advantage of raw materials produced at low cost in Korea. From the outset, they were established in Korea as an integral part of the industrial complex in Japan. In this way, these new industries became complementary to Japan's industries rather than competitive and were consistent with the policy objectives of developing Korea as a market for Japanese manufactured goods. At the same time, the establishment of new industries did not represent the growth of native industries or the transfer of resources from agriculture. On the contrary, they tended to produce abrupt disturbances to, and detrimental effects on, native industries. A brief review of the origins of the modern textile and chemical industries in Korea will further illuminate the characteristics of modern industries in the Korean economy.

The Korean Spinning and Weaving Industry was established in 1917, and production started in 1921. It was controlled by a Japanese concern (Mitsui), and the Government-General provided direct subsidies and a guarantee of 7 percent annual dividends. The main purpose of the effort was to meet the increasing demand for cotton textile products in Korea at a time of excess demand for Japan's textile goods from other nations.[3] Furthermore, the cotton production of Korean agriculture provided an excellent potential source of raw material for the industry. Since the financing, capital goods, and technology came directly from Japan, modern facilities were established both in spinning and in weaving, rendering disturbances to the rural economy, especially to the household industries which previously undertook most weaving. This may be contrasted with the growth pattern of Japanese textile industries, where modern spinning industries expanded along with small-scale weaving industries.

The origin of the modern chemical industry exhibits even more clearly the "imposed" nature of modern industries in Korea. In this connection, it should be noted that modern industries in Japan were seriously challenged by the lower cost of similar products in other nations during the 1920s. Thus Japanese entrepreneurs were forced to meet the foreign challenge by finding various ways of reducing the cost of production. In the case of the chemical industry, the cost of electric power was critical. For example, over 30 percent of the total cost in the production of ammonium sulphate was attributed to the cost of electric power.[4] During the early 1920s, a new technique of producing hydroelectric power was developed, and Korea showed a tremendous potential for hydroelectric power using the new technique. As the availability of cheap hydroelectric power became known, the Japanese chemical industries decided to extend their establishments to Korea.

It was a holding company of the Japan Nitrogen Corporation that did the pioneering work in Korea. The Korean Water Power Corporation was established in 1926 and the Korean Nitrogen Fertilizer Corporation in 1927; they were later merged. It was a holding company (100 percent) of the Japan Nitrogen Company, supported by the Mitsubishi concern. The purpose of its establishment in Korea was to reduce the cost of production, and the output was distributed to both the Korean and Japanese markets. The key success of the modern chemical industry in Korea demonstrated to Japanese entrepreneurs the feasibility of building large-scale modern industries in the Japanese colony, and the effect of this knowledge was one of the major factors contributing to the industrialization of the 1930s.

In summary, modern industries established in Korea during the 1920s had no close relationship with native industries. From the outset, they were established by Japanese capital and technology, and their major purposes were to supplement the shortages of Japanese industry or to take advantage of the low labor costs in Korea. Thus, the choice of industry and the selection of technologies were made in reference to the needs of the Japanese economy.

The Growth of Small-Scale Factories

In the earlier analysis, we pointed out that the growth rate of factory output outstripped that of the household-industry output during the late 1920s. This was achieved by the rapid expansion of new factories as shown in Table 46. The number of factories rose from 252 in 1911 to 2,384 in 1921, and to 4,914 in 1927. The number of Korean workers engaged in the manufacturing sector was substantially increased during the same period.

Table 47 presents the number of factories and the amount

Table 46
KOREAN FACTORIES
NUMBER, CAPITAL, EMPLOYEES
1911, 1921, 1927

	1911	1921	1927
Number of factories	252	2,384	4,914
Capital (K¥ 1,000)	10,614	179,143	542,646
Number of employees	14,575	49,302	89,142
Japanese	2,136	6,330	6,163
Korean	12,180	40,418	78,347
Foreigners	259	2,554	4,632

Source: The Government-General, *Statistical Year Book*, 1911, 1921, 1927.

of capital invested according to the nationality of their ownership. The major characteristics of the factories revealed in this table may be summarized as follows:

First, a "modern factory" is defined in terms of having capital of over K¥ 1 million. There were 10 to 15 of these factories until 1927, with an average capital per factory of K¥ 8,283,000. These factories were characterized by: large-scale operations, the use of modern technology and capital-intensive methods of production, the distribution of output to foreign (mainly Japanese) as well as to domestic markets, Japanese ownership, and their close relationship to the modern industries of Japan.

Second, the majority of factories owned by Japanese and other foreigners were medium- or small-scale factories. These factories were characterized by small-scale operations, the "pre-modern" technology, and their orientation was to meet exclusively the domestic demands of Korea. Given the limited size of the domestic market, the "pre-modern" type of Japanese factories were found to be profitable in Korea. They were financed largely by the accumulated capital of merchants residing in Korea, or by Japanese immigrants with modest capital, and had no financial ties with the large landlords or the financial giants (for example, Zaibasu) in Japan.

Table 47

KOREAN FACTORIES BY THE NATIONALITY OF OWNERSHIP
1921, 1927

	1921	1927
Japanese Ownership		
Modern factories (capital over K¥ 1 million)		
Number of factories	9	14
Total capital (K¥ 1,000)	40,500	115,960
Capital per factory (K¥ 1,000)	4,500	8,283
Other factories		
Number of factories	1,267	2,265
Total capital (K¥ 1,000)	153,570	303,558
Capital per factory (K¥ 1,000)	121	134
Korean Ownership		
Number of factories	1,088	2,457
Total Capital (K¥ 1,000)	7,752	23,289
Capital per factory (K¥ 1,000)	7	9
Foreign Ownership		
Number of factories	20	93
Total capital (K¥ 1,000)	13,680	18,609
Capital per factory (K¥ 1,000)	684	200

Source: The Government-General, *The Statistical Yearbook*, 1921, 1927.

Third, most factories owned by Koreans were characterized by minimum provisions of capital, and were simply an extended form of the traditional household industries which met the official definition of a factory in terms of the number of workers employed. There were no basic changes in these factories from the traditional form of native industries, and they produced mostly vegetable oils, fish fertilizers, Korean liquor, noodles, and so forth. New factories of small-scale operation were expanded in response to both the increasing domestic demand and the real-cost advantages. The latter advantages were largely provided by the development of agriculture prior to the rapid industrialization of the 1930s. Thus, over 80 percent of manufactured products in Korea used raw materials originating in the agricultural sector.[5] Only about 10 percent of the total manufactured product was accounted for by those products for which the mining sector provided raw materials.

Of special importance for our purpose is the failure of the native industries to be transformed into modern ones. On the one hand, the domestic demand for manufactured goods increased during the early years of the colonial period because of the overall growth of the economy, the rapid expansion of population, and the Japanese immigrants; on the other, the growth of native industries was checked by the rapid increase in the "pre-industrial" type of Japanese factories established by the Japanese residents in Korea and by the detrimental effects of the modern industries as already discussed. Furthermore, native industries failed to receive financial support from the financial institutions, and the large Korean landlords did not have any incentive for financing native industries. Thus, instead of developing into modern factories, they simply enlarged the scale of operation, increasing the number of workers employed, and thus were reclassified as "factories" in the official data. Therefore, the growth in the number of Korean factory workers in Table 46 reflects largely the effect of the increased workers in household industry rather than any shifts of workers from agriculture to manufacturing. For example, full-time Korean workers in manufacturing during 1921–1927 increased by only 20.4 percent according to the official estimates of year-end population by occupation, whereas the relevant figures in Table 46 show a near doubling during the same period.

The Industrialization of the 1930s

As we have seen, the spurt of industrialization did not originate from factors indigenous to the Korean economy, but was induced by basic changes in Japan's policy during the 1930s. The major objectives of the new policy during the 1930s in Japan were to replace the imports of highly fabricated products with those produced within the Japanese empire, and to develop the economy of the Japanese empire for the eventual war. The rapid increase in aggregate demand arising from the new policy provided a strong incentive on the part of Japanese entrepreneurs to develop heavy industries within the Japanese empire. It is said that the new policy was the "political solution" to "the grave problems confronting the Japanese economy in the 1920s."

> Once Japan had decided to follow a policy of military expansion, it was a relatively simple matter to step up the growth of the economy, even though the standard of living of the mass of the people did not necessarily maintain the pace. Military expenditures became gigantic in order to fight in Manchuria, Inner Mongolia, China, and in preparation for World War II. The armament industry and associated heavy industries were the leading sectors of this growth. (It did not matter at the time that this was a very artificial expansion for heavy industry, largely based on subsidies creating high-cost producers.) This was accompanied by an expansion drive to the colonies. Colonies were made to take exports, native populations were exploited as food producers, and colonial markets were frequently monopolized for the mother country's benefit.[6]

Thus the industrial structure of Japan during this time was characterized by the rapid growth of heavy industries, and the concentration of business in the hands of a small group of families, called *zaibatsu*.[7]

The industrialization of the 1930s in Korea must be viewed in context of Japan's new policy during the 1930s. It was a part of an "expansion drive to the colonies." Korea possessed certain real cost advantages for heavy industries such as an abundant capacity to generate hydroelectric power, a cheap labor force well-trained in Japanese customs and language, a strategic location in the Japanese empire, and so forth. As in Japan, the leading role in the industrialization of Korea was performed by the same group of Japanese *(zaibatsu)*, who prospered among Japanese entrepreneurs during the 1930s.

Thus, it is quite evident that the industrialization of the 1930s in Korea was largely the extension of Japan's rapidly growing industries to Korea to take advantage of the above opportunities. What should be noted here is the fact that these cost advantages in Korea were created in a sense by the new policy objectives of Japan during the 1930s. The rapid industrialization in Korea reflected neither an increased aggregate demand within the Korean economy nor any real cost advantages vis-à-vis the world market. Instead, the industrialization was a result of the over-all reallocation of resources within the Japanese empire in accordance with the policy objectives of the 1930s.

The Structure of the Manufacturing Sector

As expected, a drastic change was made in the structure of the manufacturing sector during the rapid industrialization of the 1930s. Table 48 demonstrates the direction and magnitude of changes in terms of the relative distribution of manufactured products between heavy and light industries. According to these figures, the relative share of heavy industry rose from 23.1 to 50.3 percent during the 1930s. The accelerated growth of the manufacturing sector during the 1930s was mainly confined to heavy industries and had little effect on the domestic supply of consumer goods, which continued to rely upon imports from Japan. Thus, Korean industrialization was a reversal of the normal sequence in the industrialization process, in that the major emphasis was placed on the development of heavy industries from the outset.

Closer observation of Table 48 shows that the chemical industry played the leading role in the industrialization of the 1930s. Concerning the composition of the chemical industry, Andrew J. Grajdanzev writes that "on the one hand there are numerous small enterprises with primitive techniques, making fish oil and fish fertilizers, pressing vegetable oils, and so forth; on the other hand, there are a few giant plants with the most modern techniques.[8] For example, the large factories (with over 200 workers) in the chemical industry produced over 76 percent of the total product in 1939, even though they were only 1.9 percent of the total number of factories in this industry. The composition of chemical prod-

Table 48

COMPOSITION OF FACTORY PRODUCT BETWEEN HEAVY AND LIGHT INDUSTRIES

1930-1940

(percentages)

Industry	1930		1935		1940	
Light Industry[a]	76.9		62.8		49.7	
Textiles		16.3		13.6		12.4
Food		29.2		27.8		19.9
Heavy Industry[b]	23.1		37.2		50.3	
Chemicals		15.0		24.3		37.3
Total	100.0		100.0		100.0	

Sources: Market values for factory products are from *Chōsen tōkei nempō,* 1930, 1935, 1940. The net product ratios used for the derivation of net product, on which the above figures are based, are the estimates of the Cabinet Bureau of Statistics of Japan, cited by Ohkawa in his *The Growth Rate of the Japanese Economy Since 1878*, p. 87.

a. Textiles, food, ceramics, lumber, printing, and miscellaneous products.

b. Chemicals, machines and tools, metals, gas and electrical products.

ucts in 1940 is shown in Table 49. It is quite evident in this table that the output of modern factories accounted for the major portion of chemical products.

On the whole, the type of output from the modern industries established during the 1930s predominantly consisted of intermediate goods rather than final goods, and these were destined to be used by Japan's industrial complex. In order to show the importance of foreign markets in the industrialization of Korea, the relevant data are summarized in Table 50. The figures indicate that the ratio of exported manufactured product to total production increased substantially during the 1930s. The growth rate of

Table 49

THE COMPOSITION OF CHEMICAL PRODUCTS IN 1940

Type of Goods	Market Values (1,000 K¥)	Percentage Distribution
Drugs and Medicines	9,340	1.43
Chemicals for industrial use[a]	264,008	40.69
Dyeing material and paints	238	0.05
Soap	23,041	3.55
Mineral oil[a]	40,628	6.26
Vegetable oil	6,513	1.00
Other oil and fat[a]	117,497	18.11
Rubber footwear	22,757	3.51
Fertilizers[a]	128,724	19.84
Other chemical products	36,092	5.56
Total	648,838	100.00

Source: Statistical yearbook of the Government-General, 1940.

a. Types of output where modern industries accounted for the major portion.

the exported manufactured product was even higher than the accelerated growth rate of domestic production during this period.

In this way, a strictly complementary rather than competitive relationship was maintained between the new industries in Korea and the rapidly expanding industrial complex in Japan. The emphasis on the production of intermediate goods for Japanese markets greatly weakened any relationship that might have developed between the modern industries and the small-scale industries. Thus, the modern industries rendered few linkage effects.

The rapid growth of modern industries also led to the formation of a dual structure within the manufacturing sector. The large-scale factories equipped with modern facilities and technologies had no close ties with the factories of the "pre-industrial" type, and this allowed the dual structure to prevail during this

Table 50

MANUFACTURED PRODUCT

RATIO OF EXPORTS TO DOMESTIC PRODUCTION, 1920–1941

Period	Index of Domestic Production (1)	Proportion of Exports to Domestic Production (percent) (2)
1920–1921	64.1	27.7
1924–1926	103.7	32.7
1929–1931	100.0	37.5
1934–1936	196.3	47.1
1939–1941	258.2	63.2

Sources by column:

(1) Computed from Table B–3.

(2) Computed from the Bank of Korea, *Annual Economic Review of Korea, 1948*, statistics, p. III-48, and price index of manufactured product from Table B–2.

time. Since the modern sector was geared to the market of the Japanese empire as a whole, the dual structure in the Korean economy did not retard the growth of the modern sector. Since the Government-General favored the expansion of large industries, small-scale factories and household industries were discriminated against through the various controls of the "war economy" of the late 1930s. For example, the enactment of the "Temporary Control of Capital Funds" in 1937 was instrumental in channeling capital funds exclusively into those industries connected with Japan's armament.[9]

In sum, the manufacturing sector had two distinct segments. On the one hand, it contained the most advanced sector by world standards, but this was largely a physical extension of the armament industry in Japan. On the other hand, it contained a "pre-industrial" sector consisting of both Japanese and native industries. The output of the advanced sector dominated the manufacturing

Table 51

FACTORIES AND FACTORY PRODUCTS
BY SIZE AND TYPES OF INDUSTRY, 1939
(percentages)

Types of Industry	Small Number	Factory[a] Product	Medium Number	Factory[b] Product	Large Number	Factory[c] Product
Textile	67.8	5.9	25.2	8.3	7.0	85.8
Metal	78.5	4.8	18.2	6.8	3.3	88.4
Machine & Tool	74.3	17.0	22.4	30.3	3.2	52.7
Ceramic	74.0	10.8	22.8	16.7	3.2	72.5
Chemical	78.7	9.9	19.4	14.1	1.9	76.0
Lumber	82.0	52.1	17.0	47.9	0.5	-
Printing	84.2	25.5	15.0	54.7	0.8	19.8
Food	90.9	41.6	8.5	53.2	0.6	5.2
Gas & Electricity	73.6	49.8	23.5	37.0	2.9	13.2
Others	79.9	34.8	17.5	39.8	2.6	25.4
Total	81.7	16.5	16.3	21.7	1.2	61.8

Source: Akitake Kawai, *Chōsen kōgyō no gendankai* (Seoul, 1943), p. 252-253.

a. Less than 100 workers.

b. Between 100-200 workers.

c. Over 200 workers.

sector as a whole, (see Table 51), while the "pre-industrial" sector suffered from government control and the competition of the advanced sector with respect to the product market, factor market, and so forth. The industries in the advanced sector were often characterized not only by the use of modern, capital-intensive technology, but also by a monopoly power in the Korean economy.

The "Imposed" Nature of Industrialization

As already mentioned, the growth of modern industries in Korea was achieved by the extension to Korea of Japan's industrial complex connected with armaments. Thus the industrialization of the 1930s intensified Japanese control of the manufacturing sector in the Korean economy. While most factories of Japanese ownership in Korea during the 1920s were financed by small investors, the new industries of the 1930s were established by the financial giants of Japan. According to the figures in Table 52, over 98 percent of the chemical plants, the dominant industry in the Korean economy, and over 90 percent of paid-in capital in manufacturing were owned by the Japanese.

Table 52

PERCENTAGE DISTRIBUITON OF FIRMS AND PAID-IN CAPITAL BY TYPE OF INDUSTRIES AND NATIONALITY OF OWNERSHIP
(December 1941)

Types of Industry	Number of Firms		Paid-in Capital	
	Koreans	Japanese	Koreans	Japanese
Textiles	38.5	61.5	14.56	85.44
Metal, Machine & Tools	27.8	72.2	5.73	94.27
Beverages	70.3	29.7	46.36	53.64
Pharmaceuticals	53.7	46.3	49.74	50.26
Ceramics	20.8	79.2	3.73	96.27
Rice Mills	50.9	49.1	19.65	80.35
Food	15.6	84.4	3.00	97.00
Wood	16.5	83.5	15.05	84.95
Printing	46.4	53.6	17.24	82.76
Chemicals	29.9	70.1	1.95	98.05
Others	28.3	71.7	11.65	88.35
Totals	41.4	58.6	9.10	90.90

Source: The Bank of Korea, *The Annual Economic Review of Korea, 1948*, p. 1-318.

It is not surprising that the native industries were excluded from the mainstream when we consider the industrial organization prior to the 1930s. As pointed out in the earlier analysis, the native industries suffered from competition with the "pre-industrial" type of Japanese factories in Korea and from a lack of financial resources. The growth of native industries was only a remote possibility without the positive support of the government and the institutional arrangements necessary to channel surplus funds from the large landlords to the industrial sector. In the absence of these preconditions, only a few modern industries of Japanese ownership emerged in the Korean economy prior to the 1930s.

The industrialization of the 1930s was an "imposed" development in the sense that the spurt originated in Japan's policy of expanding armaments. To implement the government policy, it was essential to expand heavy industries requiring lumpy investments. Thus, the Japanese government had to enlist major support from the financial giants in Japan, and their industrial activities were extended to Korea. Even the "pre-industrial" type of Japanese factories in Korea encountered various difficulties since the new policy was designed to reallocate productive resources in favor of modern industries.[10]

As shown in Chapter IV, the employment conditions of Korean workers were not improved, nor was there any substantial reallocation of workers from agriculture to manufacturing during the rapid industrialization of the 1930s. The modern industries were highly capital-intensive, and were built and operated with the direct importation of technology and technicians from Japan. This minimized the effects on the industrial distribution of labor.

To sum up, the rapid industrialization of the 1930s installed a dual industrial structure in the manufacturing sector of the Korean economy by extending Japan's modern sector to Korea. However, the advanced sector of the dualistic structure was largely unrelated to the indigenous factors of the Korean economy. Even the remarkable speed of the 1930s industrialization failed to harness the traditional sector toward modern economic growth.

Chapter VII

EXTERNAL SECTOR ANALYSIS

In the preceding analysis, it has often been pointed out that the Korean economy was developed as a complementary region of the Japanese economy. Thus the impetus to Korean economic growth during the colonial period came mainly from abroad through the rapid expansion of foreign transactions. The present chapter will focus on this aspect of Korean economic growth. Among the flows across national boundary,, those of human resources, commodity-products, and capital are selected for the present analysis in view of their significance, imprinted on the pattern of Korean economic growth.

Human-Resource Flows

As for the origin of Japanese immigration to Korea, H. K. Lee points out:

> The Japanese came during the Hideoyoshi invasion in 1592. Later they were allowed to reside in the three trading ports in South Kyungsang Province, within limited areas and limited as to number. By virtue of the treaty concluded between Korea and Japan in 1876, they came in greater numbers in the following decades and reached the number of 171,543 in 1910. Since the annexation, the Japanese population in Korea has been increasing more rapidly, as a result of government encouragement and private pursuit of gain.[1]

After the establishment of Japan's colonial government in Korea, the Japanese population accounted for over 2 percent of the Korean population, and showed a steady increase by immigration from Japan during the colonial period. The Korean emigration also increased, with a sudden jump during 1935–1940, as shown in Table 53.

Of great importance here are the economic functions connected with the Japanese in Korea. To illustrate this, Tables 54 and 55 are provided. While there are certain reservations about the accuracy of the year-end estimates of population, as pointed out in Chapter II, the estimates of Japanese population in Korea are believed to be fairly accurate in view of their small size and close contact with the Government-General. Among the various breakdowns of population in the official data, the category of "full-time male worker" is used in Table 55 to approximate the labor force by nationality. Let us now summarize the major findings.

First, it is quite evident that the Japanese population in Korea was increasingly concentrated in non-agricultural sectors over the period under review. During the rapid industrialization of the 1930s, the net increase of the Japanese population was found mainly in manufacturing and its related sectors, reflecting the significant role played by the Japanese population in Korean industrialization.

Second, while the Japanese population in Korea was only 2 or 3 percent of the Korean population, its labor force accounted for relatively high proportions of workers in the key sectors of the economy. Since the Japanese residents in the agricultural sector were mainly landlords, their number was very small compared to the Koreans. On the other hand, the ratio of Japanese workers to Koreans in official and professional services was extremely high, indicating that the Japanese dominated the official and professional positions. We also find relatively high ratios of Japanese workers to Koreans in manufacturing and its related sectors. This observation, along with the analysis of capital ownership,[2] suggests that the Japanese population in Korea had not only a monopoly control of the agriculture sector, but also provided a major portion of the labor force trained for industrialization.

Finally, it may be noted that the Japanese immigration of skilled workers discouraged the training of Korean workers. Instead, the unskilled Korean workers emigrated. This may be seen by comparing the quality of workers involved in the Japanese immigration and the Korean emigration during the 1930s. As

Table 53

POPULATION AND ITS EXTERNAL FLOWS

KOREAN AND JAPANESE, 1920-1940

Annual Average	Population[a] (1,000s)		Flow (1,000s)		Percentage	Percentage
	Korean	Japanese	Korean emigration[b]	Japanese immigration[c]	(3) to (1)	(4) to (2)
	(1)	(2)	(3)	(4)	(5)	(6)
1920-1925	18,142	395	31	89	0.2	22.5
1925-1930	19,729	485	56	68	0.3	14.0
1930-1935	21,323	573	81	69	0.4	12.0
1935-1940	22,878	664	205	56	8.9	8.4

a. The average of the beginning and terminal years, census data.
b. From Kin Tetsu, *Kankoku no jinkō to keizai* (Tokyo, 1965), p. 36.
c. Chōsen kosei kyokai, *Chōsen ni okeru jinko ni kansuru sho tokei* (Seoul, 1943).

Table 54

INDEX OF JAPANESE RESIDENTS IN KOREA BY OCCUPATION

	Total	Agr.	Fishery	M-sector[a]	Commerce Transp.	Official & Professional Services	Others	Unemployed & Unreported
1920	100	100	100	100	100	100	100	100
	(348)[b]	(40)	(11)	(60)	(117)	(102)	(13)	(5)
1925	122	98	118	112	114	138	162	200
1930	144	105	118	122	134	174	246	380
1935	168	93	91	135	150	232	178	420
1940	198	83	91	242	163	253	223	480

Source: Computed from Chōsen Kōsei Kyōkai, *Chōsen ni okeru jinkō ni kansuru sho tōkei* (Seoul, 1943), pp. 25–26.

a. Including mining sector.
b. Number of persons in 1,000.

Table 55

FULL-TIME MALE WORKERS
(in 1,000 workers)

	Manufacturing			Commerce and Transportation			Official and Professional Service			National Average
	Korean	Japanese	Ratio J/K	Korean	Japanese	Ratio J/K	Korean	Japanese	Ratio J/K	Ratio J/K
1925	102	20	20%	281	37	13%	101	45	45%	-
1930	107	23	22	296	40	14	117	52	44	3.0%
1935	138	25	19	312	42	13	140	62	44	-
1940	159	32	20	378	44	12	148	70	47	3.5%

Sources: Official estimate, 1920 (Statistical Yearbooks of the Government-General, 1920). Population Census data, 1925, 1930, 1935, 1940.

pointed out, most of the Japanese immigrants were absorbed by the non-agricultural sectors that were closely related to the industrialization of the 1930s. In contrast, most of the Korean emigrants were unskilled manual workers with little training for industry.

We may now place in perspective the implications of these human-resource flows. The Japanese population in Korea exerted a major influence on Korean economic growth by occupying the strategic positions. From the early years of the colonial period, the Japanese residents in Korea dominated the administrative setup and, to a lesser degree, commerce and industry. The supply of Japanese workers in Korea was favorably affected by the dualistic structure of the labor market in the Japanese economy. Beginning from the early 1900s, Japan developed a differential structure which led to the creation of an abundant supply of labor in the traditional sectors.[3] It was from this reservoir of labor that the task of colonial development could be carried out by immigration of Japanese. This, in turn, severely restricted the opportunities for technological diffusion and institutional changes conducive to modern economic growth by Koreans.

Commodity Flows

The quantitative data on foreign trade constitute one of the oldest series available in Korea.[4] However, the data cover only the flow of commodity-products and gold transactions.[5] It should also be noted with respect to the trade statistics of the colonial period that complete economic integration of Korea with the Japanese Empire made it extremely difficult to account for all private transactions, especially goods carried out by Japanese residents in Korea. Thus, the official data on foreign trade during the colonial period tend to underestimate the actual magnitude of flows. Nevertheless, the commodity-trade statistics are the only data available for the entire period under review, and some broad trends in the external flows of goods may be derived from an analysis of the limited data at hand.

Let us begin with the scale of over-all growth in commodity flows. This may be observed in terms of the proportion of com-

Table 56

FOREIGN TRADE RATIOS
1911-1940
(percentages based on values at current prices)

Annual Averages	Export[a] Ratio	Import[b] Ratio	Trade[c] Ratio
1911-1915	6.6	13.4	20.0
1916-1920	12.2	14.8	27.0
1921-1925	18.6	19.2	37.8
1926-1930	22.5	26.2	48.7
1931-1935	24.9	27.3	52.2
1936-1940	25.8	34.6	60.4

Sources: Trade data are taken from the Bank of Korea, *Annual Economic Review of Korea, 1948*, statistical data, III, 43. Total commodity-products are taken from Appendix A.

a. Percentage of export to total commodity-product plus import.
b. Percentage of import to total commodity-product plus import.
c. Export ratio plus import ratio.

modity trade to total availability of products in the economy (domestic production plus imports). The trade ratios are presented in Table 56. The figures show a substantial increase throughout the entire period. The growth rate of commodity trade was much higher than the growth rate of the total available product. Furthermore, the rapid industrialization of the 1930s increased the dependency of the national economy on external transactions.

The rapid increase in export ratios during the early years of the colonial period should be noted. Until the late 1920s, the growth rate of exports was higher than the import rate. This corresponds to a period of rapid agricultural development, resulting in Korea's becoming the major source of food supply to Japan. After the late 1920s, however, the growth rate of imports outstripped that of exports. This was a significant change in the pattern of trade caused by the industrialization of the 1930s. The industrialization of the 1930s was not directed toward the development of import-substituting industries. Instead, it increased the import ratios by increasing the imports of heavy machines and

Table 57

RELATIONSHIP OF FOREIGN COMMODITY TRADE TO SIZE OF COUNTRY
1938-1939

Countries Arranged in Descending Order of Population Size	Number of Countries	1938-1939 Average Population (millions)	Average Foreign Trade Ratio
I	10	135.4	0.17
II	10	16.2	0.24
III	10	7.3	0.31
IV	10	3.7	0.38
V	12	1.5	0.38
Korea	1	23.1	0.53[a] 0.56[b]

Source: Figures except Korean are from Simon Kuznets, *Six Lectures on Economic Growth*, p. 96.

a. Assuming that commodity-product accounts for 65% of GNP.
b. Assuming that commodity-product accounts for 75% of GNP.

tools. The export ratios of the 1930s also included the transit trade to Manchuria originating from Japan.

With the limited scope of our estimates borne in mind, it may be desirable to compare them with the trade ratios of other nations as shown in Table 57.[6] While we recognize the differences in the scope and reliability of the data, the comparison nevertheless leads to the conclusion that the trade ratios of the Korean economy during the colonial period were exceptionally high compared to the international averages. The high ratios point up the fact that foreign trade was the main channel in developing the Korean economy as an integral part of the Japanese Empire.

The percentage composition of international commodity flows according to type of goods is presented in Table 58. The figures are derived from the official trade statistics at current prices. The crude foodstuffs include all food grains, and the manufactured foodstuffs are the output of food industries. Crude materials for industries consist of various raw materials, and the distinction

Table 58

PERCENTAGE DISTRIBUTION OF FOREIGN TRADE COMMODITIES BY TYPE OF GOODS, 1915-1941
(based on current values)

Annual Average	Crude Materials			Manufactured Products			
	Food	Indus-trial	Total	Food	Semi-finish-ed	Finished	Total
				EXPORTS			
1915-1919	48.7	20.5	69.2	1.0	7.6	22.2	30.8
1920-1924	63.0	11.1	74.1	1.7	11.0	13.2	25.9
1925-1929	64.2	8.7	72.9	2.5	12.2	12.4	27.1
1930-1934	59.1	15.6	74.7	3.8	15.6	5.9	25.3
1935-1939	41.8	15.2	57.0	3.5	21.0	18.5	43.0
				IMPORTS			
1915-1919	7.0	6.4	13.4	7.1	14.9	64.6	86.6
1920-1924	11.6	14.7	26.3	5.9	20.2	47.6	73.7
1925-1929	18.3	11.7	30.0	5.8	17.1	47.1	70.0
1930-1934	10.3	21.9	32.2	6.7	12.6	48.5	67.8
1935-1939	7.2	13.5	20.7	4.6	13.2	61.5	79.3

Source: Computed from the Bank of Korea, *Annual Economic Review of Korea, 1948*, Statistical Data, III, 50-59.

between semi-finished and finished manufactured goods is made literally with respect to their readiness for final use.

The major trends of the commodity structure, as revealed in Table 58, may be explained in relation to the structural changes in the Korean economy that took place during the colonial period. As for the composition of export products, the relative share of crude materials showed an initial rise corresponding to the early years of agricultural development, and then remained at a high level of over 70 percent until the middle 1930s. When the rapid industrialization of the 1930s took place, however, the relative share of crude materials was substantially reduced, whereas the share of semi-finished and finished manufactured products showed a sharp increase.[7]

The major trends in the structure of imported commodities also agree with our expectations based on the preceding analysis. While the Korean economy was mainly an exporter of agricultural products, imports of crude food stuffs increased at a time of rapid agricultural development, in order to meet the increasing shortage in the domestic food supply by imports of inferior grains. As mentioned, the food shortage in Korea was created by the increasing portion of domestic food production that was channeled into the Japanese market, coupled with the rapid growth in population.

The relative share of manufactured products was initially over 86 percent of the total amount of goods imported, and then showed a declining trend until the middle 1930s. When Korea was annexed to Japan, most manufactured goods were imported. However, as the factories (mostly small-scale) expanded during the 1920s following the repeal of the Corporation Law, some of the domestic demand for manufactured products was met by their production. This was especially true of food, textiles, and other industries which benefited from the abundant supply of raw materials provided by the rural sector. The industrialization of the 1930s required the import of machinery and tools without reducing the import of consumer goods, thereby substantially increasing the realtive share of finished manufactured goods.

Table 59 presents the percentage distribution of foreign commodity trade according to the countries of origin and destination. As expected, the figures point to trade domination by Japan and China (mainly the *yen* bloc of Manchuria and a part of the Chinese mainland). The close relationship of foreign trade to the structure of the Korean economy may be shown by comparing the pattern of trade with Japan to Korea's trade with other nations. The relevant data for this purpose are summarized in Tables 60 and 61.

When the Korean economy was developed as the major supplier of food grains and raw materials to Japan prior to the industrialization of the 1930s, Korean exports were increasingly directed towards Japan, with food grains the major export product. The industrialization of the 1930s, however, increased the relative

Table 59

PERCENTAGE DISTRIBUTION OF FOREIGN TRADE BY ORIGIN AND DESTINATION
1910-1941
(based on current values)

Annual Average	Japan	China	U.S.	Others	Total
		EXPORTS			
1910-1912	73.7	16.8	2.3	7.2	100.0
1919-1921	89.9	8.8	0.1	1.2	100.0
1929-1931	91.4	8.2	0.1	0.3	100.0
1939-1941	77.4	21.2	0.2	1.2	100.0
		IMPORTS			
1910-1912	62.4	10.1	8.6	18.9	100.0
1919-1921	64.4	23.7	7.7	4.2	100.0
1929-1931	76.5	15.9	2.2	5.4	100.0
1939-1941	88.3	4.8	1.6	5.3	100.0

Sources: 1910-1921, *Chōsen tōkei nempō.*
1929-1941, The Bank of Korea, *Annual Economic Review.* 1948, III, 44-45.

share of manufactured products, especially the share of unfinished products, reflecting the nature of Korean industrialization as complementary to the Japanese economy. At the same time, Korean export of manufactured products to other parts of the *yen* bloc increased during the 1930s, following Japan's territorial expansion toward China. It should be noted, however, that this increase did not entirely represent Korean manufactured goods, because an increasing portion of total exports was accounted for by manufactured products made in Japan. The Manchurian Incident of 1931 placed Korea in a strategic location for trade between Japan and China. Thus a rapid expansion of transit trade took place during the 1930s.

The national origin of imported goods also reflects the

Table 60

COMPOSITION OF GOODS TRADED WITH JAPAN
(percentages based on current values)

Year	Crude Materials			Manufactured Products				Miscellaneous Products
	Food	Industrial	Total	Food	Semi-Finished	Finished	Total	
				EXPORTS				
1910	69.5	20.0	89.5	-	0.4	3.6	4.0	6.5
1919	72.0	13.2	85.2	0.4	9.7	0.3	10.4	4.4
1929	68.9	14.1	83.0	2.5	8.8	2.0	13.3	3.7
1935	57.2	16.5	73.7	1.6	17.2	4.1	22.9	3.4
1939	27.9	21.6	49.5	2.5	32.4	10.9	45.8	4.7
				IMPORTS				
1910	3.3	8.6	11.9	11.8	10.3	54.2	76.3	11.8
1919	1.6	5.2	6.8	5.1	9.3	64.7	79.1	14.1
1929	9.8	3.5	13.3	8.1	10.4	61.0	79.5	7.2
1935	6.0	10.3	16.3	6.6	14.2	58.8	79.6	4.1
1939	5.7	8.8	14.5	5.1	14.4	61.4	80.9	4.6

Source: Computed from the Bank of Korea, *Annual Economic Review, 1948*, III, 50–59.

Table 61

COMPOSITION OF GOODS TRADED WITH NATIONS OTHER THAN JAPAN

(percentages based on current values)

Year	Crude Materials			Manufactured Products				Miscellaneous Products
	Food	Industrial	Total	Food	Semi-Finished	Finished	Total	
				EXPORTS				
1910	56.2	16.4	72.6	0.4	0.9	2.4	3.7	23.7
1919	30.4	11.9	42.3	14.5	2.8	7.2	24.5	33.2
1929	19.7	12.8	32.5	12.5	9.2	19.6	41.3	26.2
1935	15.6	10.6	26.2	8.6	14.9	39.3	62.8	11.0
1939	21.3	6.1	27.4	7.2	4.1	54.2	65.5	7.1
				IMPORTS				
1910	3.4	13.7	17.1	9.9	10.0	60.0	79.9	3.0
1919	20.8	28.3	49.1	8.3	6.4	27.0	41.7	9.2
1929	30.6	20.1	50.7	6.6	14.2	26.1	46.9	2.4
1935	34.7	30.7	65.4	4.3	11.3	15.0	30.6	4.0
1939	30.2	44.2	74.4	1.8	10.5	9.3	21.6	4.0

Source: See Table 60.

Table 62

BALANCE OF COMMODITY TRADE
1911–1940
(million K¥ at 1936 prices)

Annual Average	Net Balance With Japan	Net Balance With Other Nations	Total Balance
1911–1915	-24	-29	-53
1916–1920	4	-28	-24
1921–1925	52	-59	-7
1926–1930	23	-72	-49
1931–1935	-12	-37	-49
1936–1940	-209	9	-200

Source: Statistical yearbooks of the Government-General.

changing structure of the Korean economy. During the early years of the colonial period, Korean imports were mainly finished manufactured goods from both Japan and other nations. However, there was increasing concentration on imports from Japan as the colonial period progressed. As for imports from other countries, crude materials gradually dominated the scene because of Korea's shortage of domestic foodstuffs and the industrialization of the 1930s. For this reason, a complete monopolization of Korean imports by Japan was unattainable.

Finally, let us examine the balance of commodity trade, as displayed in Table 62. According to these figures, the foreign trade of Korea was characterized by over-all deficits throughout the entire period under review, indicating net additions to the goods available for domestic use. While trade with Japan produced an export surplus during 1918–1932 (except 1929, 1930), it was outstripped by deficits in trade with other nations. The large deficit during the early years of the colonial period was caused largely by the import requirements of the colonial government for an efficient administration, and the sharp increase of deficits during the later years should be attributed to the imports of machines and tools for the industrialization of the 1930s. On the other hand, the trade surplus with Japan was realized through rice exports.

Capital Flows

Here we are primarily interested in the magnitude of capital flows from Japan to Korea during the colonial period. At the outset, however, we must distinguish between real capital transfer and financial transfer. The former refers to the movement of goods and services, while the latter simply indicates the transfer of purchasing power. The very success of Japanese colonial policy in developing the Korean economy as an integral part of the Japanese Empire made financial transfer the most common type of capital flow between the two countries. However, its magnitude is impossible to measure because some private capital flows were simply not reported, and the distinction between the two countries was often omitted in the official data concerning financial transfers.

Fortunately, what is important for our purpose is the real capital transfer, and its magnitude may be approximated from the foreign trade statistics. While the official trade statistics suffer from the omission of services, it was our assumption in the preceding section that the rough magnitude of international flow of goods could be estimated from the trade data. The net inflow may be approximated as the deficit of commodity trade less the export surplus of gold.

Table 63 presents the magnitudes of real capital inflows during the colonial period. The domestic need for foreign capital in connection with the colonial administration and the industrialization of the 1930s, plus the nation's capacity for gold exports, were the factors determining the variations in real capital inflow.

When the annual averages of real capital inflows in Table 63 are used to estimate the total amount of capital inflow for the period as a whole, its amount reaches K¥ 1,403 million during 1910–1941.[8] This may be compared with the total amount of Japanese capital invested in Korea during 1910–1941. To the best of our knowledge, there was only one estimate available of Japanese capital invested in Korea as of 1941, made by the Seoul Chamber of Commerce,[9] and shown in Table 64. According to this estimate, the total amount of Japanese capital invested was K¥ 7,329 million during 1910–1941. This amount includes both

Table 63

ESTIMATES OF REAL CAPITAL FLOWS
BETWEEN KOREA AND JAPAN
1911-1940
(million K¥ at 1936 prices)

Annual Average	Net Balance of Commodity Trade	Net Balance of Gold Transaction	Net Inflow
1911-1915	-53	17	36
1916-1920	-24	8	16
1921-1925	- 7	4	3
1926-1930	-49	7	42
1931-1935	-49	45	4
1936-1940	-200	58[a]	142

Sources: Table 62.
Statistical yearbooks of the Government-General.

a. Since the data for gold transactions are not available from 1937 on, the 1936 figure is taken as the average of 1936-1940.

the real capital inflow from Japan and reinvestment of profits in Korea. The former is approximated by the balance of payments account in Table 63 so that the relative share of real capital inflow from Japan in total capital investment by Japanese in Korea is about 20 percent. Thus, it may very well be that over 80 percent of Japanese capital invested in Korea during 1910–1941 originated within the Korean economy.

In short, the present analysis shows that the magnitude of real capital inflow from Japan was only a small portion of total capital owned by the Japanese in Korea. This suggests that a substantial proportion of total capital formation was financed within the Korean economy but controlled by the Japanese during the colonial period.

Table 64

JAPANESE-OWNED CAPITAL INVESTED IN KOREA

(cumulative totals in million K¥, 1910-1941)

Government funds	2,071
Loans of banking system	342
Corporate investments	3,941
Private funds	973
Others	2
Total	7,329

Source: Seoul Chamber of Commerce, *Chōsen ni okeru naiji shihon no toka genkyō* (Seoul, 1944), pp. 39-42.

Chapter VIII

REGIONAL PATTERN OF ECONOMIC GROWTH SOUTH VS. NORTH

The economic growth of Korea during the colonial period was accompanied by broad regional specialization. It is well known that the south of Korea became a mainly agricultural region, while the north was developed as an industrial region. In order to observe the regional development in some detail, the present chapter re-examines the pattern of Korean economic growth in terms of the division between the south and the north of Korea.[1]

The present analysis will also contribute to an understanding of the initial conditions for the postwar economic growth of a divided Korea. There were at least two major changes during the postwar period that transformed the Korean economy so as to make the preceding period distinct from the postwar period in the long-term study of the Korean economy. These were the termination of the colonial period and the partition of the economy between the south and the north.

The original purpose of the partition of Korea was purely military, designed as a transitory setup "to divide the responsibilities of accepting surrender (Japan) rather than to set up zones of military occupation."[2] Unfortunately, however, it turned out to be an enduring situation, thereby creating two independent regimes hostile to each other. Thus, the analysis of regional development as well as the over-all growth of the Korean economy during the colonial period is essential for understanding the initial conditions for postwar economic growth in both the south and the north.

Population Distribution

Let us first examine population growth in the south of Korea in comparison with that in the north during the colonial period. In Table 65, population census data are classified between south and

Table 65

GROWTH OF TOTAL POPULATION
IN SOUTH AND NORTH KOREA
1925-1944
(thousands of persons)

Korean Population	1925	1930	1935	1940	1944
All Korea					
Population	19,020	20,438	22,208	23,547	25,133
Net Increase		1,418	1,770	1,339	1,586
Percentage Increase		7.4	8.6	6.0	6.7
The South[a]					
Population	13,005	13,900	15,020	15,627	16,574
Net Increase		895	1,120	607	947
Percentage Increase		6.8	8.0	4.0	6.1
The North[a]					
Population	6,014	6,537	7,187	7,920	8,558
Net Increase		523	650	733	638
Percentage Increase		8.7	9.9	10.0	8.0

Source: Computed from Chōsen Sōtokufu (Government-General of Korea), *Chōsen kokusei chōsa hōkoku,* 1925, 1930, 1935, 1940, 1944. Populations for the south and the north of Korea are derived by totaling the populations of provinces belonging to each of these two regions.

a. See note 1 (Chapter VII) for the coverage of north and south Korea.

north. According to these figures, the growth rate of population in the north had always been higher than in the south. While the average growth rate of the entire Korean population per decade during 1925–1944 was about 15 percent, the corresponding rate for the south was 13 percent and for the north, 19 percent. What were the major causes for the differential in growth rates between the two regions?

Reliable data concerning birth and death rates, internal migration, and emigration by provinces are nonexistent. Therefore some indirect methods have to be used to explain the differential rate of population growth.

We begin by assuming identical death rates for the two regions during the colonial period. This may be justified because the major portion of the reduction in death rates during this period was brought about by improved public health and these benefits were spread throughout the country. If the death rates were the same in the two regions, the differential growth rates of population might be explained by the differences in birth rates and internal migration moving from the south to the north.[3]

To present a comparison of birth rates in the two regions, the numbers of children aged 0–4 per 1,000 women of ages 15–49 are presented in Table 66. Assuming the same death rate (including infant death), the figures indicate that birth rates in the north were higher than in the south during the early years of the period under review. However, the differences in birth rates were reduced substantially during the later years, in contrast to the differential growth rates of total population which persisted throughout the entire period. Therefore, regional differences in birth rates cannot fully account for different north/south growth rates of total population.[4]

If we rule out the differences in the vital rates as the main cause of differences in growth rates of total population between the two regions, there remains the effect of internal migrations from the populous areas of the south. It was shown in Chapter V that some farmers tended to leave their land during the later years of the colonial period to escape increasing burdens. This trend must have been accompanied by internal shifts of workers from rural areas of the south to the industrializing north, particularly in the mining and manufacturing sectors of the region. It may very well be that internal migration relieved some of the population pressure in the densely populated areas of the south.

Table 66

NUMBER OF CHILDREN (0-4)
PER 1,000 WOMEN (15-49), BY PROVINCES, 1925-1940

	1925	1930	1935	1940
All Korea	706	696	726	731
South Korea				
Kyunggi do	660	633	704	696
Choongchung bookdo	676	703	735	768
Choongchung namdo	696	698	754	778
Chunla bookdo	684	703	702	714
Chunla numdo	687	700	709	713
Kyungsang bookdo	725	704	725	750
Kyungsang namdo	736	719	727	710
Kangwon do[a]	684	699	729	758
Arithmetic means	694	695	723	736
North Korea				
Whanghai do	694	663	744	744
Hamkyung bookdo	749	717	712	710
Hamkyung namdo	753	720	730	730
Pyungan bookdo	721	709	736	729
Pyungan namdo	736	701	751	745
Arithmetic means	731	703	735	731

Source: "Korea and the Koreans in the Northeast Asian Region," *Population Index* (October 1950), p. 283.

a. About one-half of this province belongs to North Korea.

As for the degree of urbanization, the census data showing the distribution of population between rural and urban areas are summarized in Table 67. According to these figures, the share of the urban population was larger in the south than in the north during the period under review. However, the pace of urban concentration was accelerated in the north during the late 1930s' industrialization. The rapid increase of urban population in the north seems to support our previous conclusion that the internal migration of population from the south was absorbed mainly by

Table 67

POPULATION DISTRIBUTION: URBAN VS. RURAL
1925–1949
(thousands of persons)

Area	1925	1930	1935	1940	Percentage change 1925–1940
The South					
Urban[a]	655	840	1,942	1,950	298%
Rural	12,042	12,332	13,496	13,269	110%
Ratio of urban to rural population (%)	5.4	6.8	14.3	14.7	
The North					
Urban[a]	195	350	464	868	445%
Rural	6,631	7,537	7,797	8,239	124%
Ratio of urban to rural population (%)	2.9	4.6	5.9	10.5	

Source: Computed from *Chōsen kokusei chōsa hōkoku,* 1925, 1930, 1935, 1940.

a. Urban area refers to a *fu* (city) with population of 25,000 or over.

the manufacturing centers of the north.

Commodity-Production

We shall attempt to analyze the comparative characteristics of south and north Korea with respect to the level and industrial composition of commodity-product. However, there is one serious limitation, namely, that agricultural and forestry products cannot be classified by province until 1934.[5] Accordingly, our analysis of total commodity-product between the two regions uses as bench marks the interval of 1934–1935 through 1939–1940, which reflects the phase of rapid industrialization.

Table 68

RATIOS OF CURRENT VALUES OF COMMODITY-PRODUCT BETWEEN SOUTH AND NORTH KOREA BY INDUSTRIAL ORIGIN
1934–1935 and 1939–1940

	1934		1935		1939		1940	
	South	North	South	North	South	North	South	North
Agriculture	.661	.339	.640	.360	.558	.442	.632	.368
Forestry	.547	.453	.553	.447	.542	.458	.515	.485
Fishery	.634	.366	.653	.347	.642	.358	.622	.378
Mining[a]	.242	.758	.242	.758	.242	.758	.242	.758
Manufacturing	.630	.370	.552	.448	.458	.542	.466	.534

Source: Computed from the official data on "Manufactured Commodities" (*Chōsen tōkei nempō*, 1934, 1935, 1939, 1940), classified by the south and north of Korea. (See note 1).

a. Mining available by 1934.

The following methods are used to derive net values of commodity-product originating in the south and the north. First, the current market values of products by industrial origin are classified according to place of production, and ratios between the two regions are derived as shown in Table 68. Second, the ratios of Table 68 are applied to net commodity-product at the constant prices of 1936 (Table A-12) to derive the net values by region. These values are presented in Table 69. While our estimates are crude, they may provide an empirical basis from which broad inferences may be derived concerning the comparative characteristics of the two regions during the colonial period. Let us summarize our major findings.

First, the relative shares of commodity-product (Table 68) in the two regions show generally stable values during the period under review except in manufacturing, where the figures indicate a basic change during the process of rapid industrialization.[6] While

Table 69

COMMODITY-PRODUCT BY INDUSTRIAL ORIGIN AND REGION
1934-1940
(million K¥ at 1936 market prices)

Industry	Annual Averages			
	The South		The North	
	1934-35	1939-40	1934-35	1939-40
Agriculture	646.9	496.3	348.1	333.3
Forestry	58.6	65.8	48.0	58.7
Fishery	44.5	80.2	24.6	46.8
Mining	15.6	28.9	48.9	90.6
Manufacturing	125.9	157.9	89.1	183.7
Total	891.5	829.1	558.7	713.1

Sources: Computed from Tables 68 and A-12. The relative shares of commodity-product in the two regions were applied to the estimates of net commodity-product.

the south was producing about 63 percent of total manufactured products during 1934, the share was reduced to 47 percent by 1940. Thus, the north outran the south in producing manufactured products during the last half of the 1930s.[7]

Second, the net commodity-product originating in the south (Table 69) shows a slight decline during 1934/35–1939/40, reflecting the agrarian character of the economy.[8] However, the net commodity-product in the north was expanded by 27.6 percent during the same period as a result of the rapid industrialization. The contrasting trends between the two regions lead us to conclude that the industrialization of the 1930s in Korea was mainly confined to the north. Accordingly, the northern share of national commodity-product increased from 38 percent to 46 percent during the period under review.

Third, the level of per capita commodity-product in the two regions shows different trends during the interval of 1934–1935 through 1939–1940. The relevant data are summarized in

Table 70

TOTAL AND PER CAPITA COMMODITY-PRODUCT AND POPULATION, 1939–1940
(annual averages by region)

Region & Year	Population (1000's)	Total Commodity-Product (million K¥)	Per Capita Commodity-Product (K¥)
The South			
1934–1935	15,020	891.5	59.4
1939–1940	15,627	829.1	53.1
The North			
1934–1935	7,187	558.7	77.3
1939–1940	7,920	713.1	90.0

Sources: Computed from Tables 65 and 69.

Table 70. According to these figures, the level of per capita commodity-product in the south was considerably lower than the corresponding level in the north, reflecting the north's more favorable ratio of population to commodity-production. While per capita commodity-product in the south shows once again a slight decline, due mainly to the extremely poor harvest in 1939, that originating in the north shows a substantial increase. This comparison should not be taken to represent the per capita income differences between the two regions because the service sector, which probably was concentrated in the south, is not included in our estimates.

We may now examine the industrial composition of commodity-product, as shown in Table 71. The figures indicate that over 72 percent of total commodity-product in the south originated in agriculture and 14 percent in manufacturing during 1934–1935. In the north, however, agricultural product was about 62 percent of commodity-product as a whole versus 16 percent for manufactured product. Thus, there was no substantial

Table 71

INDUSTRIAL COMPOSITION OF COMMODITY-PRODUCT
1934–1940
(percentages by region)

Industry	The South 1934–1935	The South 1939–1940	The North 1934–1935	The North 1939–1940
Agriculture	72.6	59.9	62.3	46.8
Forestry	6.6	7.9	8.6	8.2
Fishery	5.0	9.7	4.4	6.5
Mining	1.7	3.5	8.8	12.7
Manufacturing	14.1	19.0	15.9	25.8
Total	100.0	100.0	100.0	100.0

Source: Computed from Table 69.

difference between the two regions in the industrial composition of commodity-product during the early phase of industrialization.

The structural differences between the two regions were widened by rapid industrialization during the later years of the colonial period. For example, the relative share of agricultural product in the south during 1939–1940 was about 60 percent, whereas the corresponding share in the north was reduced to 47 percent. On the other hand, the percentage share of manufactured products in the north showed an increase of 10 percent, while in the south the share of manufactured product showed only a moderate increase of about 5 percent. The structural changes accompanying the industrialization in Korea made it legitimate to call the south an agricultural region and the north a manufacturing region. The industrialization of the colonial period brought a development of regional specialization in the Korean economy.

A supplementary relationship between the two regions in commodity-product during the colonial period may be further demonstrated by observing regional differences in the composition of output within the key sectors of the economy—agriculture and manufacturing. In the absence of the relevant data, the composition of agricultural product between the two regions may be

Table 72

PERCENTAGE DISTRIBUTION OF CULTIVATED AREAS BETWEEN SOUTH AND NORTH KOREA FOR FOOD GRAINS IN SELECTED YEARS

	1910	1922	1936
Rice			
South	73	72	74
North	27	28	26
Summer Grains			
South	73	73	73
North	27	27	27
Beans			
South	47	47	42
North	53	53	58
Cereals[a]			
South	13	20	20
North	87	80	80
Others[b]			
South	48	56	60
North	52	44	40

Source: Computed from Kobayakawa, Statistical Appendix, Table III-9.

a. Italian millet, barnyard millet, glutinous millet, sorghum, corn, buckwheat.

b. White potatoes, sweet potatoes, vegetables.

approximated by the percentage distribution of cultivated areas by type of food grains, as shown in Table 72. The figures indicate that the south was highly specialized in the production of rice and barley, while the north produced about half or more of the remaining food grains. In this connection, it may be recalled from the preceding analysis that the pattern of food consumption during the colonial period was deliberately shifted by substituting other food grains for rice, so as to export the maximum amount of rice. Therefore, the north played an important role in the

Table 73

NET VALUE OF MANUFACTURED COMMODITIES IN NORTH AND SOUTH KOREA DURING 1930-1931 AND 1939-1940

	1930-1931 (two-year totals)			1939-1940 (two-year totals)		
	(1) South	(2) North	(3) (1)/(2)	(4) South	(5) North	(6) (4)/(5)
	(in million K¥)			(in million K¥)		
Textiles	11.1	2.9	3.79	64.1	12.9	4.95
Metals	1.8	0.8	2.33	6.0	49.6	.12
Machines & Tools	7.1	1.5	4.67	44.7	17.2	2.60
Ceramics	3.4	8.6	.40	17.4	47.1	.37
Chemicals	13.9	19.9	.70	80.6	399.2	.20
Wood Products	2.6	0.8	3.07	10.4	8.2	1.27
Printing	9.0	0.7	12.65	18.5	2.9	6.33
Foods	30.7	15.1	2.03	126.2	70.6	1.79
Others	28.6	13.0	2.20	101.1	40.0	2.53
Total	108.2	63.3	1.71	469.0	647.7	.72

Note: Ratios of "manufactured commodities" in south and north Korea, at producers' current prices, are applied to net values of manufactured commodities. The net values are computed from market values by using the "net product ratios" of Table 4.

domestic supply of food grains, while the south produced the major portion of agricultural products exported to Japan.

The rapid industrialization of the 1930s meant not only different growth rates of manufacturing between the south and the north, but also some basic changes in the composition of output, particularly in the north. In order to make a regional comparison of these changes, the net values of manufactured products for selected years are classified by type of industries in Table 73. According to these figures, the south was still producing almost twice the value of manufactured products as the north during 1930-1931. The north showed a predominance only in the

Table 74

PERCENTAGE DISTRIBUTION OF MANUFACTURED PRODUCT BY TYPE OF PRODUCT AND REGION
(1930-1931 and 1939-1940)

Type of Product	1930-1931		1939-1940	
	South	North	South	North
Textiles	10.2	4.6	13.7	2.0
Metals	1.7	1.2	1.3	7.7
Machinery & Tools	6.5	2.4	9.5	2.7
Ceramics	3.2	13.5	3.7	7.3
Chemicals	12.9	31.4	17.2	61.6
Wood Products	2.3	1.4	2.2	1.3
Printing	8.4	1.2	3.8	0.3
Foods	28.4	23.8	26.9	10.9
Others	26.4	20.5	21.7	6.2
Total	100.0	100.0	100.0	100.0

Source: Computed from Table 73.

production of ceramics and chemicals. However, figures for 1939–1940 indicate that the total value of manufactured product originating in the north outweighed the value of the south, mainly reflecting the rapid expansion of chemical and metal products. This resulted in basic changes in the composition of manufactured products in the north, as shown in Table 74. For example, the percentage share of producers' goods[9] in the north during the decade of the 1930s increased from 35 percent to 72 percent, whereas in the south the corresponding figures showed only a moderate increase, from 21 percent to 28 percent.[10] Thus the north was completely transformed into a producer of producers' goods during the 1930s, while the south maintained its heavy orientation toward the production of consumer goods.[11]

Chapter IX

KOREAN ECONOMIC GROWTH IN PERSPECTIVE

Let us now summarize the major characteristics of Korea's colonial growth experience. For this purpose, Kuznets's study on the uniformity of the historical changes in the process of economic growth is very helpful.[1]

Kuznets defines "modern economic growth" as a substantial and sustained rise in total and per capita product, and its characteristics may be seen in "new patterns of population growth, industrialization, urbanization, new patterns of use in national product, an increase in non-personal forms of economic organization and a rise in relative importance of achievement in the social values.[2]

Modern economic growth, which has a marked degree of uniformity, is normally achieved by increasing the productivity of existing resources as well as the expansion of resources. As prerequisites for growth, some institutional changes and technological development are required in the pre-industrial economy. Since the degree and effect of technological development and institutional change are not evenly distributed in all sectors of the economy, the initial development efforts tend to be concentrated in a few sectors, based on the comparative advantages and the preferences of society. The disequilibrium forces thus created gradually yield to "chain effects" of development, thereby ensuring a substantial and sustained rise in national income. All of these changes leading to modern economic growth are deeply influenced and accelerated by international contacts. Accordingly, the characteristics of modern economic growth are also revealed in new patterns of external transactions.

The "Imposed" Nature of Colonial Development

In many respects, the growth pattern of the Korean economy during the colonial period displays the characteristics of modern economic growth as described above. The colonial period witnessed a substantial rise in total and per capita commodity product, rapid expansion of population and urbanization, substantial changes in economic structure through the rapid industrialization of the 1930s, and a remarkable expansion of foreign trade.

However, the similarity of the growth pattern in Korea to that of modern economic growth elsewhere is only superficial because of the "imposed" nature of the growth in the Korean experience; the growth was not carried out by "those spontaneous forces of growth in society that arise from ordinary men and women."[3] As shown in Chapters I and III, economic growth during the colonial period was achieved under the rigorous leadership of the colonial government, backed by a strong military and police presence. In order to implement the policy objectives of colonial rule, new institutions were created and old ones abolished by a series of reforms beginning in the transitional period. The institutional reforms placed the privileged group (Japanese residents in Korea and some Korean landlords) in a strategic position to achieve their policy objectives. While various incentive measures were introduced for the privileged group, the majority of the Korean population was mobilized through coercive measures.

The major objectives of economic development policy in Korea were derived from Japan's overall economic circumstances: her growing food shortage, the need for foreign markets for her expanding industrial production, and her factor endowment. Policies were carried out with much more vigor and force in Korea than in Japan for the simple reason that colonies could be readily coerced by power. Accordingly, the individual incentive of an average inhabitant was seriously impeded because the colonial rule did not provide any real connection between his efforts and rewards during the process of rapid development, as shown in the striking contrast between domestic production and consumption.

The pattern of colonial economic development was also characterized by a lack of "chain effects" following from the unbalanced growth of the Korean economy. The rapid industrialization during the colonial period was largely the product of Japanese capital, technology, and entrepreneurs. The development was designed to serve the needs of the Japanese Empire, as shown by the extremely high foreign trade ratios during this period. As a result, the modern sectors of the Korean economy were simply irrelevant to the traditional sectors and the majority of inhabitants.

Dualism in Comparative Perspective

The characteristics of colonial development may be further elaborated in terms of the dualism created during the colonial period. It should be noted at the outset that dualism in the process of economic growth is not unique to the Korean economy of 1910–1940. Japan is a well-known case of industrial dualism created during the early decades of modern economic growth.[4] Therefore, our analysis will be focused on the factors that made the Korean dualism (colonial type) different from that of Japan.

Since the agricultural and manufacturing sectors dominated the Korean economy, the analysis will deal mainly with those two sectors. Let us begin with the productivity differences. As shown in Table 75, the ratio of real product per worker in agriculture to that in manufacturing declined from 0.91 to 0.24 during 1920–1940. In the case of Japan, the productivity ratio changed from 0.50 in 1886–1898 to 0.23 in 1931–1938.[5] The comparison clearly illustrates the remarkably rapid widening of the gap between the two sectors. How was this achieved?

As already noted, the industrialization of the 1930s was heavily concentrated in producer-goods industries (including the production of unfinished goods) in response to the demand created by Japan's policy toward armament. These industries required modern technologies of a capital-intensive nature, minimizing the labor requirements that might have created large-scale interindustry shifts from the agricultural sector. In addition to this technological reason, the over-all factor endowment of the Japanese economy also favored capital-intensive industries in Korea. It is well known that the factor proportions in the Japanese economy of the 1930s shifted from labor surplus to capital abundance. Given the relatively free mobility of factors between the two countries and the lack of an adequate labor force for modern industries

Table 75

REAL PRODUCT PER WORKER IN AGRICULTURE AND MANUFACTURING
1920–1940
(in *yen* at 1936 prices)

Year	Agriculture (A)	Manufacturing (M)	M–A	A/M (percentage)
1920	137	150	13	91
1925	134	212	78	63
1930	148	305	157	49
1940	170	699	529	24

Sources: Gainful workers—year-end estimates, taken from statistical yearbooks of the Government-General.
Real commodity-product—Table A-12.

in Korea, it is not surprising to find that advanced technologies with highly trained engineers were directly imported from Japan during the 1930s. In short, the 1930s' rapid industrialization was carried out by foreign capital, technology, and personnel, and in response to a demand exogenous to domestic conditions.[6] Accordingly, the development failed to transform the traditional sectors into highly productive ones; it only accelerated the widening productivity gap between the two sectors.

When there is a widening productivity gap between agriculture and manufacturing, one would normally expect the gap to be counterbalanced by terms-of-trade effects. However, as shown in Table 76, the counterbalance did not materialize in the Korean economy. Contrary to expectation, the terms of trade moved against agricultural products during the industrialization of the 1930s. This can be explained in terms of the following characteristics of Korean economic growth during the colonial period.

First, the land-tenure system of the colonial period forced native farmers to channel their shares of domestic production into the market, thus acting as suppliers of agricultural products. It should be noted here that the products thus marketed did not

Table 76

TERMS OF TRADE BETWEEN AGRICULTURAL AND MANUFACTURED PRODUCTS, 1911-1940 (1936=100)

Annual Average	Agricultural Price Index (A)	Manufacturing Price Index (M)	A/M (percent)
1911-1915	52	62	83.9
1916-1920	104	129	80.6
1921-1925	114	142	80.3
1926-1930	96	110	87.3
1931-1935	79	93	84.9
1936-1940	125	149	83.9

Source: Table A-11.

represent a voluntary "surplus" over consumption. On the other hand, the subsidiary activities of farm-households in producing non-agricultural products were severely curtailed by the economic policy of the Government-General, thus increasing the demand for manufactured goods.

Second, the industrialization of the 1930s had very little effect on the industrial distribution of Korean workers. It thus failed to create any substantial degree of urbanization which might have increased the demand for agricultural products at a rate much faster than the actual growth rate. On the other hand, the lack of other employment opportunities forced the agricultural sector to accept an extremely low level of compensation relative to the modern sectors.

Third, the industrialization of the 1930s was largely concentrated on the production of non-consumer goods. Thus, instead of increasing the supply of manufactured consumer goods in the Korean economy, the industrialization responded mainly to Japan's policy of armament during the 1930s.

To the sectoral dualism thus mentioned should be added our earlier findings that a dual structure was developed within each sector. For example, in manufacturing there was a division into two exclusive sections: modern industries catering to markets abroad and small-scale (or handicraft) industries to meet demands from the majority of the population. Even in the latter case, the Japanese residents in Korea expanded their scope of operations at the expense of Korean establishments. In agriculture, large landlords with irrigation facilities accounted for over 60 percent of total rice exports.

Dualism in the Korean economy may be examined from another angle. We turn now to the pattern of income distribution by factor shares during the period under review. In the absence of relevant data for any rigorous analysis, only some broad trends can be observed. Out of the total commodity-product, payments are made as compensation to employees, income of entrepreneurs, and returns to assets used in production, and so forth. For our purpose, we shall examine the difference in the pattern of income distribution between the compensation of workers (including tenant farmers) and the income of entrepreneurs and asset-owners. The former group encompasses the mass of inhabitants, while the latter consists of landlords and industrialists, including most of the Japanese residents in Korea. The factor shares in non-commodity sectors are excluded from the present analysis.

Let us first consider the trend in real wages of gainful workers. In this regard, the index of wages in Seoul is the only series available covering the entire period under review.[7] To obtain the index of real wages, the wage index of Seoul is deflated by the wholesale price index (the only available official data), and the results are summarized in Table 77 along with the indices of real product per worker in agriculture and manufacturing.

It should be noted that the wage rate in Table 77 is very likely to be higher than the average wage rate of rural areas (including the income of tenant farmers), because the former represents some skilled workers, and urban wage rates are usually higher than rural wages to attract workers from rural areas. This observation is confirmed by comparing the real-wage index to indices of product per worker in agriculture and in manufacturing. On the whole, the

Table 77

INDICES OF REAL WAGES AND PRODUCT PER WORKER

Annual Averages	Price Index (1)	Wage Index: Money Wage (2)	Wage Index: Real Wage (3)	Index of Product Per Worker: Agriculture (4)	Index of Product Per Worker: Manufacturing (5)
1918–1922	140	155	100	100	100
1928–1932	99	199	128	108	203
1938–1942	170	145	94	124	466

Sources by Column:

(1) Table A-11, General Price Index.
(2) The Bank of Korea, *Annual Economic Review of Korea*, 1949, IV, 426–435 (1910=100).
(3) Computed from columns (1) and (2).
(4) & (5) Table 75 (1920, 1930, 1940 figures).

figures indicate a substantial decrease in real wages during the industrialization of the 1930s, whereas the product per worker showed a substantial gain, particularly in manufacturing.

We observed in the earlier analysis that there was a trend for workers to leave rural areas. In this connection, it is generally held that, in a typically backward country where agriculture traditionally provides the major source of livelihood and where family ties are very strong, "only strong incentives could attract inhabitants to urban districts, especially in family or household units."[8] However, it is quite evident here that the "strong incentives" for leaving agriculture did not come from increasing real wages in the non-agricultural sectors. Instead, the labor migration was motivated by the extreme deterioration of rural life during the late colonial period.

The exodus of workers from rural areas in turn reinforced the low level of wages in non-agricultural sectors by engendering an excess supply of workers. Let us quantify the magnitude of this surplus of labor. We note that there was a net increase in the male population of 1,449,000 during 1930–1940. If we assume the activity rate of 1930 to prevail during 1940, the net addition to the total supply of gainful workers would have been 868,000, whereas the actual increase of employment was only 215,000 (see Table 21). Thus, it may well be said that a surplus of about 653,000 workers was created while the rapid industrialization was taking place in Korea during 1930–1940. In other words, the industrialization failed to provide sufficient employment opportunities for those who were driven out of the agricultural sector.

Accordingly, many workers who left rural areas went to stay in urban areas, involving themselves in the "miscellaneous" category of extremly low productivity. To a certain extent, they were absorbed into the mining sector. Unlike the manufacturing sector, however, the expansion of the mining sector (arising from the development of heavy industries) was largely carried out by manual workers to take advantage of cheap labor. The number of workers employed in this sector was only a very small portion of the labor surplus, as estimated above. This labor surplus was also the main reason for the accelerated growth of emigration during the decade.

The insufficient employment opportunities for the inhabitants was once again evident in the declining trend of female workers during 1930–1940. To be sure, the reduction may be at least partly explained by the effect of urbanization; the separation from rural life and handicraft industries curtailed their employment opportunities. However, the determining factor for this drastic decline in female workers during the decade must have been the increasing surplus of unskilled workers in the economy during the process of rapid industrialization and population growth. Industrialization of a capital-intensive nature failed to provide enough employment opportunities for the workers withdrawn from agriculture and from household industries, as well as for the net increase arising from the rapid population growth. When an excess in the labor force exists, it is normal to find labor of inferior quality driven out of the market.

In short, it may be asserted that the growing dual structure in production caused substantial reduction of real wages at a time of rapid increase in the level of product per worker. Given the declining trends in the relative share of agricultural products grown by tenant farmers and in the real wages of non-agricultural sectors, the rapid growth in total and per capita commodity-product, and the net real capital inflow, there must have been a substantial rise in the relative shares of property income and participating income of entrepreneurs and professionals. In the light of this consideration, it seems evident that the colonial development and rapid industrialization in Korea caused a trend toward increasing inequality in size distribution of income, and this in turn created a higher level of savings.

Some fragmentary data are available to support this conclusion. For example, deposits at commercial banks, listed by the nationality of their owners during the 1930s, show that over 70 percent of the total savings in Korea were owned by Japanese residing there, who numbered only 2.6 percent of the total population.[9] This may be explained by their special positions in the Korean economy, as reflected in the industrial distribution of Japanese workers in Korea.[10]

Dual structure by nationality in the Korean economy may be seen in the differentiated training of labor force through formal education. Prior to Korea's annexation to Japan, formal education was available only to the ruling class of society (*yangban*), and it was carried out on a tutorial basis (called *sodang*) at private homes. However, the colonial administration introduced the new educational system which prevailed in Japan. Primary education (elementary schools), secondary education (high schools), and higher education (colleges and universities) constituted the new educational system. Table 78 summarizes data on the graduates of the new educational system for selected years.

The introduction of the new educational system broke the monopoly by the ruling class, and formal education was open to the public for the first time in Korean history. However, as the figures in Table 78 indicate, the spread of higher education was very limited. Only elementary education showed a success in this respect. Formal education was also limited among the Koreans

Table 78

TOTAL NUMBER OF SCHOOL GRADUATES IN KOREA
(selected years)

Year	Graduates[a] of Elementary School	Graduates[a] of High School	Graduates[a] of Higher Institutions
1915	14,556 (0.09)[b]	1,348 (0.00)	69
1920	16,845 (0.10)	1,899 (0.01)	134
1925	52,326 (0.27)	5,301 (0.03)	378
1930	77,308 (0.37)	7,127 (0.03)	726
1935	128,254 (0.56)	10,831 (0.05)	974

Source: Computed from *Chōsen tokei nempō* 1915, 1920, 1925, 1930, 1935.

a. Includes both Koreans and Japanese in Korea.
b. Figures in parentheses are percentages of the total population.

when compared with the Japanese in Korea and with those in Japan proper. These comparisons are made in Table 79.

In summary, certain features of dualism were similar between Korea and Japan. In both cases, the dual structure stemmed from development strategies designed to accelerate the economic objectives of government. Dualism in the labor market rendered low wage rates at a time of rapid productivity increase. Changes in income distribution favored big enterprises. Financial and nonfinancial institutions alike contributed to both increasing savings (voluntary and forced) and channeling them into modern industries.

The fundamental difference in the dualism of the two countries, however, lay in the coincidence of economic dualism with different nationalities. Unlike a typical colonial policy of Western nations which developed export-oriented enclaves, Japan attempted a complete integration of the Korean economy with Japan's. The key role was played by a minority economic elite consisting mostly of Japanese residents in Korea. As a result, all aspects of the Korean economy were linked with the Japanese economy; modern

Table 79

NUMBER OF STUDENTS IN ALL TYPES OF SCHOOLS PER THOUSAND OF POPULATION IN KOREA AND JAPAN

	In Korea 1939		In Japan
	Korean Students per 1,000 of Korean Population	Japanese Students per 1,000 of Japanese Population	1936
Primary Schools	55.2	142.8	164.5
High Schools	1.31	32.7	17.9
Professional Schools	1.18	12.1	6.2
Colleges & Teachers' Seminaries	0.27	7.2	1.28
Universities	0.0093	1.06	1.03
Total	58.0	195.9	190.9

Source: Andrew J. Gradjanzev, *Modern Korea*, p. 264.

industries were linked with the modern sector of Japan, and the traditional non-agricultural sectors were dominated by Japanese residents in Korea.

Such a dual structure resulted in a social cleavage between the economic elite and the majority of the inhabitants. The two communities were never amalgamated during the period under review. In this regard, we find contrasting patterns of occupational distribution and demographic trends between the two communities.[11] Under these circumstances, the social benefits of industrialization—such as the training of skilled workers, the supply of entrepreneurs, changing habits of consumption and savings conducive to sustained economic growth, and the modernization of social values consistent with the requirements of industrialization (such as the will to economize, and so forth)—were largely confined to the elite group. On the other hand, the majority of inhabitants continued to pursue a traditional way of life, except that their productive activities were forced to expand.

Dualism in Long-Term Perspective

As for the impact of colonial development on the Korean economy in long-term perspective, let us begin with possible contributions to growth. Since the colonial regime adopted modern facilities and new institutions to implement economic policies, some concomitants of modern economic growth during the postwar period were inherited from the colonial period. Among these were, perhaps, transportation and communication facilities, a modern banking and monetary system, modern factories, educational and administrative facilities, and the various surveys undertaken relating to the potential of the Korean economy.

In addition to the modern elements created during the colonial period, the overall growth of the economy clearly demonstrated the potentials of the Korean economy for rapid economic growth and industrialization. This "demonstration effect" must have been a contributing factor to the postwar economic growth in Korea. Even though the colonial policy prevented the diffusion of modern technology among the Koreans, the "demonstration effect" was often instrumental for a limited diffusion of new knowledge to the inhabitants. In agriculture, the diffusion of new knowledge on the method of cultivation and new inputs took place quite extensively. All of these changes in the process of colonial development created favorable factors for economic growth when Korea obtained political independence at the end of World War II.

The above factors alone, however, were not sufficient to generate sustained growth during the postwar period. In order to ensure a process of continuous economic growth, much more than favorable factors in a static sense is required. High on this list are human elements such as the quality of the minority that is to assume the leadership of economic development, social values as reflected in individual behavior, and the organizational framework that will create the incentives—the will to develop, and the will to unify the population. Our study shows that Japanese colonial development failed to create these dynamic factors of economic development.

As mentioned, the minority group that assumed the leadership of the colonial development was alien to the endogenous elements of the Korean economy. The dualism in economic and social structures prohibited any internal mechanisms by which the widening gap between the two segments of the economy might have been reversed in the course of sustained economic growth so long as the Korean economy remained a periphery of the Japanese economy.

The majority of inhabitants in Korea were largely deprived of any benefits resulting from the over-all growth of the Korean economy as a whole. While their productive activities were forced to expand, the traditional pattern of the social structure was devoid of any modern transformation, as exemplified by the rural structure. The various incentive schemes and institutions of modern capitalism failed to reach beyond the minority group. Accordingly, internal frictions—between landlords and tenants, civil servants and the majority of inhabitants, and so forth—were aggravated as the dual pattern of economic growth proceeded. Only the military and police forces of the colonial regime could hold the population together. To these factors should be added the deteriorating effects of the colonial development on the scale of social values and on the individual's outlook toward modernization. The fact that the advanced sector of the dual economic structure was always identified with the colonial ruling class augmented the ideological impediments to modernization.

A rapid growth of over-all output without accompanying modernization of social structure and organizational framework is bound to create a total exhaustion of both population and resources. The imposed nature of the over-all growth and the very success of the colonial economic policy point up an enervated state of the Korean economy at the end of the colonial period, thereby rendering the postwar economic growth more difficult.

Thus, both the static and dynamic factors of modern economic growth should be considered in ascertaining the impact of colonial development on postwar economic growth in Korea. Needless to say, the termination of the colonial period in Korea meant a sudden breakdown of the major sources of colonial development and thus a shift in the economic growth pattern. Fundamental changes in the structure of the Korean economy were required during the postwar period, consistent with the endogenous factors and the international division of labor. It would be a highly rewarding task to investigate in some detail the type of adjustments required for sustained growth of the Korean economy after the colonial period. However, the partition of the country between south and north during the postwar period makes it futile to speculate on the postwar adjustments of the Korean economy as a whole.

To the extent that the colonial development had far-reaching effects on postwar economic growth, the period constitutes an integral part of Korean economic history in relation to modernization. We already ruled out the validity of the argument that the colonial period witnessed the beginning of modern economic growth. However, this study suggests that the colonial period represents a portion of Korea's transition period to modern economic growth, because the colonial development produced both the contributing and impeding factors affecting postwar economic growth. As mentioned at the outset of this study, a transition period refers to a period of intense struggle between growth-promoting forces and growth-retarding forces. In the final analysis, a successful journey from the transition period to modern economic growth depends upon the presence of the dynamic factors of economic development. Since colonial development failed to create these dynamic factors, Korea's transition period to modern economic growth had to be extended beyond the colonial period.

Appendix A

Table A-1

MARKET VALUES OF AGRICULTURAL AND FORESTRY PRODUCTS
(in current prices, K¥ million)

	Agriculture Crops	Agricultural Non-Crops	Forestry Products	Total
1910	200	21	19	240
1911	297	33	20	350
1912	366	37	20	423
1913	427	45	22	494
1914	370	46	23	439
1915	310	65	23	398
1916	387	74	24	485
1917	555	79	26	660
1918	905	104	28	1,037
1919	1081	154	29	1,264
1920	1197	129	30	1,356
1921	826	132	57	1,015
1922	928	124	73	1,125
1923	902	131	77	1,110
1924	1020	163	74	1,257
1925	1036	193	53	1,282
1926	978	185	60	1,223
1927	974	191	64	1,229
1928	842	211	74	1,127
1929	809	211	74	1,094
1930	571	165	63	799
1931	601	156	59	816
1932	718	176	55	949
1933	767	216	94	1,077
1934	905	216	106	1,227
1935	1031	221	114	1,366
1936	977	232	118	1,327
1937	1304	257	139	1,700
1938	1296	279	157	1,732
1939	1312	333	193	1,838
1940	1665	387	237	2,289

Sources: *Chōsen tōkei nempō*, except rice adjustment as per Table 1.

Table A-2

NET PRODUCT RATIOS OF RICE PRODUCTION AND RELATED INDICES
1910-1941
(base year 1933=100)

Year	Index of Fertilizer per Paddy Field	Price Index of Commercial Fertilizers	Index of Rice Yield per Paddy Field	Price Index of Brown Rice	Net Product Ratio of Rice Production
1910	n.a.	n.a.	60.7	49.3	90.00
1911	n.a.	n.a.	65.3	65.8	90.00
1912	n.a.	n.a.	60.5	82.6	90.00
1913	n.a.	n.a.	65.6	84.7	90.00
1914	n.a.	n.a.	75.2	61.6	90.00
1915	6.9	55.6	67.7	50.4	88.47
1916	8.8	40.0	72.4	61.0	88.92
1917	14.0	22.5	70.6	87.5	89.31
1918	17.5	81.7	78.0	138.9	88.22
1919	22.6	121.7	65.2	203.0	87.19
1920	19.8	156.0	75.5	197.2	87.21
1921	17.4	82.8	73.8	134.9	88.04
1922	18.8	82.4	76.0	154.5	88.22
1923	19.3	85.2	77.2	143.4	87.99
1924	40.3	82.1	70.9	174.4	86.38
1925	47.1	107.3	75.9	192.9	85.34
1926	55.5	108.5	80.4	166.7	83.94
1927	59.3	108.4	88.1	149.7	83.43
1928	73.5	110.5	83.3	129.6	79.85
1929	73.7	122.9	77.8	134.5	78.31
1930	80.6	100.0	95.3	114.5	80.02
1931	82.0	77.4	90.0	75.3	77.35
1932	91.8	74.7	94.4	101.6	80.35
1933	100.0	100.0	100.0	100.0	76.50
1934	114.6	105.0	95.9	113.5	75.08
1935	135.2	115.1	104.9	138.5	75.54
1936	143.6	132.6	95.7	145.2	71.50
1937	149.7	126.8	129.0	150.2	76.77
1938	153.4	142.8	109.9	158.0	72.96

Table A-2 (continued)

1939	209.5	184.4	191.7	180.3	58.45[a]
1940	159.4	204.2	105.3	216.4	70.72
1941	162.2	206.6	119.4	208.3	71.82

Sources: 1910–1936; *Chōsen keizai nempō,* 1939, Statistical Appendix
1937–1938; *Chōsen keizai nempō,* 1940, Statistical Appendix
1939–1941; *Annual Economic Review of Korea*, 1948, Statistics

a. The figure reflects the worst crop failure during the period.

Table A-3

NET VALUE OF AGRICULTURAL AND FORESTRY PRODUCTS, 1910–1940 (in million K¥, current prices)

Year	Crops	Non-Crops	Forestry Products	Total
1910	181	18	17	216
1911	267	30	18	315
1912	330	34	18	382
1913	384	41	20	445
1914	333	41	21	395
1915	275	59	21	355
1916	344	66	22	432
1917	496	71	23	590
1918	799	94	26	919
1919	942	138	26	1106
1920	1044	116	27	1187
1921	727	119	51	897
1922	818	111	66	995
1923	794	118	69	981
1924	881	147	67	1095
1925	884	174	48	1106
1926	821	167	54	1042
1927	790	172	58	1020
1928	672	190	58	920
1929	634	190	67	891
1930	457	149	57	663
1931	465	141	53	659
1932	577	159	50	786
1933	587	195	85	867
1934	680	195	95	970
1935	779	199	103	1081
1936	698	209	106	1013
1937	1001	321	125	1447
1938	945	251	141	1337
1939	767	299	173	1239
1940	1178	349	213	1740
1941	1303	473	310	2086

Sources: Calculated from Tables A-1 and A-2.

Table A-4

VALUE OF MANUFACTURED PRODUCTS BY GOVERNMENT-OWNED ENTERPRISES
(in million K¥)

Year	Cigarettes	Ginseng	Measuring Instruments	Other Products	Total
1910	-	-	-	0.4	0.4
1911	-	-	0.2	1.9	2.1
1912	-	0.4	0.2	1.9	2.5
1913	-	0.7	0.2	2.0	2.9
1914	-	1.3	0.1	2.4	3.8
1915	-	1.4	0.2	2.8	4.4
1916	-	1.7	0.2	4.1	6.0
1917	-	2.1	0.3	7.6	10.0
1918	-	2.0	0.3	17.1	19.4
1919	-	2.1	0.4	16.3	18.8
1920	-	2.5	0.2	18.0	20.7
1921	13.6	2.1	0.3	21.9	37.9
1922	14.0	2.3	0.3	24.0	40.6

Source: *Chōsen keizai nempō*, 1939, Statistical Appendix p. 44–45.

Table A-5

MARKET VALUES OF HOUSEHOLD PRODUCTS, 1933–1939

Year	Total Amount (in million K¥)	Percentage of Total "Manufactured Commodities"
1931	93.5	37
1932	129.0	41
1933	147.1	40
1934	167.1	38
1935	200.1	33
1936	227.8	31
1937	260.2	27
1938	281.7	25
1939	328.6	22

Sources: Data for 1933–1935 are from Keijo Shōkō Kaigi-sho, *Chōsen ni okeru gatenkogyo chōsa*.
Data for 1936–1939 are from Akitake Kawai, ibid., p. 235.

Table A-6

MARKET VALUES OF FACTORY OUTPUT BY SUBDIVISIONS

1911 - 1940

(in million K¥)

Year	Textiles	Metals	Machines & Tools	Ceramics	Chemicals	Wood Products
1911	0.3	1.0	0.2	0.3	0.5	0.1
1912	0.5	1.6	0.2	0.4	0.7	0.2
1913	1.0	1.6	0.3	0.6	1.8	0.2
1914	0.8	2.1	0.2	0.4	0.8	0.2
1915	1.9	7.3	0.2	0.5	0.4	0.5
1916	2.9	12.1	0.2	0.6	0.6	0.6
1917	6.9	22.8	0.3	0.8	0.8	1.4
1918	8.9	18.7	2.7	1.6	1.7	2.0
1919	13.8	16.4	3.5	2.7	2.8	7.7
1920	11.2	12.8	2.7	2.2	2.3	6.1
1921	12.5	21.3	4.3	5.1	2.2	5.7
1922	18.8	14.9	2.2	3.6	8.8	7.6
1923	22.3	17.7	2.7	4.3	10.4	9.0
1924	21.1	16.5	3.2	5.5	6.7	7.1
1925	19.2	24.9	3.7	7.1	11.4	10.9
1926	24.0	21.2	5.0	3.4	16.1	13.1
1927	25.9	22.2	8.6	6.3	13.0	13.9
1928	30.0	26.1	7.3	3.6	17.2	14.5
1929	38.4	20.4	6.3	9.2	17.8	8.1
1930	33.8	15.3	5.0	8.5	25.0	7.3
1931	24.5	16.1	3.6	7.4	32.3	6.6
1932	30.8	21.5	3.6	7.7	35.8	7.0
1933	38.9	29.2	4.2	8.8	52.6	10.2
1934	49.9	41.3	6.6	10.1	69.0	11.9
1935	71.4	21.3	8.5	15.2	119.2	14.8
1936	90.6	28.4	9.6	19.3	163.9	19.7
1937	123.0	45.3	10.9	21.5	269.0	27.3
1938	156.6	86.9	21.5	31.7	319.9	30.3
1939	193.7	131.7	48.6	36.3	462.0	42.7
1940	231.3	144.4	71.4	52.8	653.0	79.8

Table A-6 (continued)

Food & Beverages	Others	Rice-Mill	Printing	Gas & Electricity	Total
0.7	5.2	11.6	0.6	1.0	21.5
1.5	6.9	17.4	1.0	1.5	31.9
2.7	7.1	20.9	1.1	1.6	38.9
3.4	7.6	18.2	1.0	1.7	36.4
3.6	11.8	21.1	1.0	2.0	50.3
4.2	14.6	25.6	1.3	2.3	65.0
5.9	18.8	47.4	1.5	2.3	108.9
9.0	36.6	89.3	2.6	3.1	176.2
11.4	53.3	128.5	2.9	1.2	244.2
9.9	47.4	102.2	2.3	0.9	200.0
8.9	37.7	94.9	8.7	3.0	204.3
19.1	16.7	101.5	4.0	7.1	204.3
22.7	19.8	120.7	4.8	8.4	242.8
29.3	31.7	159.5	5.1	8.3	294.0
31.8	33.6	180.4	8.1	5.6	336.7
32.8	45.9	192.2	7.5	4.7	365.9
36.5	41.3	187.2	8.2	6.6	369.7
44.6	49.4	182.5	11.1	6.3	392.6
42.9	52.9	180.6	10.0	16.4	403.0
36.2	45.9	115.9	8.2	6.4	307.5
31.1	21.9	125.5	8.8	16.1	293.9
39.5	25.5	152.6	9.7	11.1	344.8
46.8	22.0	154.6	10.0	11.0	388.3
59.4	16.3	200.0	11.3	12.8	488.6
74.1	57.0	251.8	12.8	39.8	685.9
88.9	60.7	231.7	13.1	40.0	765.9
112.2	69.0	281.3	16.4	40.1	1,016.0
143.5	98.3	304.1	17.0	-	1,209.8
177.8	126.7	274.7	19.4	-	1,513.6
203.4	181.1	100.5	19.2	-	1,736.9

Source: *Chōsen keizai nempō.*

Table A-7

MARKET AND NET VALUES OF MANUFACTURED PRODUCT
1911-1940
(in million K¥ at current prices)

Year	Household Market Values	Household Net Values	Factory Market Values	Factory Net Values	Total Market Values	Total Net Values
1911	22	9	22	4	44	13
1912	12	5	32	6	44	11
1913	20	8	39	8	59	16
1914	19	8	37	7	56	15
1915	38	15	50	10	88	25
1916	39	16	65	14	104	30
1917	57	23	109	20	166	43
1918	91	36	176	33	267	69
1919	147	59	244	45	391	104
1920	130	52	200	37	330	89
1921	127	51	204	42	331	93
1922	133	53	204	39	337	92
1923	139	55	243	46	382	101
1924	151	60	293	53	444	113
1925	158	63	337	63	495	126
1926	169	67	366	68	535	135
1927	161	65	370	72	531	137
1928	158	63	393	79	551	142
1929	151	60	403	85	554	145
1930	115	46	307	70	422	116
1931	94	37	294	67	388	104
1932	129	52	345	74	474	126
1933	147	59	389	87	536	146
1934	167	67	489	107	656	174
1935	200	80	686	167	886	247
1936	228	91	766	199	994	290
1937	260	104	1,016	269	1,276	373
1938	282	113	1,210	317	1,492	430
1939	329	131	1,514	427	1,843	558
1940	344	138	1,737	558	2,081	696

Source: See Chapter II, pp. 22-27.

Table A-8

MARKET AND NET VALUES OF FISHERY PRODUCT
1910-1941
(in million K¥)

Year	Market Values: Marine Product (1)	Cultured Product (2)	Processed Product (3)	Total (4)	Total Net Values (5)	Ratio of (5) to (4) (6)
1910	-	-	-	7.9	5.5	69.9
1911	-	-	-	9.0	6.4	70.0
1912	8.5	-	4.6	13.1	7.5	56.9
1913	11.5	-	5.4	16.9	10.0	58.8
1914	12.1	-	6.9	19.0	10.8	57.3
1915	13.2	-	7.8	21.0	12.0	57.0
1916	16.0	-	9.8	25.8	14.5	56.6
1917	20.9	-	13.2	34.1	19.3	56.4
1918	32.9	0.1	19.1	52.1	30.0	57.1
1919	43.8	0.3	28.1	72.2	40.7	56.3
1920	39.3	0.4	21.4	61.1	35.2	57.7
1921	45.0	0.7	25.7	71.4	40.9	57.3
1922	47.5	0.6	26.4	74.5	42.9	57.3
1923	51.7	1.5	29.6	82.8	47.5	57.3
1924	52.0	1.7	31.2	84.9	48.3	56.9
1925	51.6	2.2	32.1	85.9	48.6	56.7
1926	53.7	2.5	34.1	90.3	51.1	56.5
1927	64.1	2.5	40.3	106.9	60.5	56.5
1928	66.1	3.3	44.9	114.3	64.0	56.0
1929	65.3	2.7	44.8	112.8	63.1	55.9
1930	50.1	2.4	30.4	82.9	47.1	56.9
1931	46.6	2.6	28.4	77.6	44.1	56.9
1932	46.3	2.4	27.4	76.1	43.4	57.1
1933	51.4	2.9	35.6	89.9	50.2	55.8
1934	57.8	2.8	45.5	106.1	58.1	54.7
1935	66.0	2.9	65.0	133.9	70.7	52.8
1936	79.9	4.7	79.3	163.9	86.5	52.8
1937	89.9	4.6	93.4	187.9	98.4	52.4
1938	87.1	5.9	96.8	189.8	98.4	51.8
1939	151.1	8.3	167.9	327.3	169.5	51.8
1940	175.5	15.5	181.8	372.8	227.0	60.9
1941	166.8	18.5	172.6	357.9	188.2	52.3

Sources: Data for 1910-1911 are from *Nihon keizai tokeishu*, p. 359.
Data for 1912-1936 are from *Chōsen keizai nempō*, 1959, Statistical Appendix p. 11.
Data for 1937-1941 are from *Chōsen keizai nempō*, 1943, pp. 92-99.

Table A-9

MARKET AND NET VALUES OF MINING PRODUCTS
1910–1941
(in million K¥)

Year	Mining Market	Products Net
1910	6.1	4.9
1911	6.2	4.9
1912	6.8	5.5
1913	8.2	6.6
1914	8.5	6.8
1915	10.5	8.4
1916	14.1	11.3
1917	17.1	13.6
1918	30.8	24.7
1919	25.4	20.3
1920	24.2	19.4
1921	15.5	12.4
1922	14.5	11.6
1923	17.3	13.9
1924	19.2	15.3
1925	20.9	16.7
1926	24.1	19.3
1927	24.4	19.5
1928	26.4	21.1
1929	26.5	21.2
1930	24.7	19.8
1931	21.6	17.3
1932	33.7	27.0
1933	48.3	38.6
1934	69.1	55.3
1935	88.0	70.4
1936	110.4	88.3
1937	142.0	113.6
1938	190.1	152.1
1939	230.4	184.3
1940	319.7	255.7
1941	369.1	295.3

Sources: 1910–1936, *Chōsen tōkei nempō*, 1939, Statistical Appendix, pp. 14–15.
1937–1941, Nihon Tōkei Kenkyūjō, *Nihon keizei tōkeishū*, pp. 359–60.

Table A-10

VOLUME INDICES OF AGRICULTURAL CROPS
1929-1931=100

Year	Rice	Barley	Beans	Other Grains	Cotton	Total
	(1)	(2)	(3)	(4)	(5)	(6)
1910	64.03	63.01	68.78	61.31	14.30	63.6
1911	71.18	74.76	79.68	68.90	18.09	72.0
1912	66.85	78.50	89.56	72.68	23.44	71.0
1913	74.51	90.07	91.30	89.45	26.69	80.3
1914	86.95	82.21	92.53	82.04	26.76	85.7
1915	79.04	89.61	98.85	90.14	32.32	84.1
1916	80.19	87.39	104.77	98.39	32.25	86.7
1917	84.22	92.45	107.66	102.48	48.98	90.8
1918	94.11	109.50	123.39	112.69	52.78	102.7
1919	78.19	94.41	73.62	73.33	65.99	78.9
1920	91.57	100.08	118.39	117.27	77.71	100.0
1921	88.14	103.31	113.13	116.85	64.63	97.4
1922	92.38	93.72	106.64	103.16	80.42	96.5
1923	93.37	81.77	110.79	103.10	86.44	95.9
1924	81.34	98.45	84.22	97.79	93.42	87.2
1925	90.90	105.75	109.80	94.01	94.98	94.7
1926	94.15	97.34	105.37	94.84	109.82	95.5
1927	106.44	92.15	113.87	97.66	102.98	102.9
1928	83.14	88.76	90.24	100.45	115.78	86.7
1929	84.31	95.28	94.87	104.13	107.18	89.1
1930	118.02	101.12	106.40	104.98	114.36	113.9
1931	97.66	103.60	98.61	90.88	78.52	98.3
1932	100.57	107.70	104.54	104.28	104.53	103.6
1933	111.94	105.25	107.98	93.49	107.99	109.2
1934	102.86	112.82	93.28	71.63	105.01	100.8
1935	110.04	124.94	105.52	89.24	144.78	111.8
1936	119.43	105.60	90.23	93.95	93.08	111.8
1937	164.88	148.99	102.40	105.85	162.80	151.4
1938	148.53	119.35	93.06	93.50	142.52	134.1
1939	88.30	133.34	55.10	90.44	142.50	93.3
1940	132.45	126.91	125.43	76.74	126.04	125.8
1941	153.12	116.77	71.94	54.86	138.37	128.9

Sources: Columns (1)-(5): 1910-37, calculated from *Chōsen keizai nempō*, 1939, Statistical Appendix, pp. 6-7.; 1938-41, calculated from *Annual Economic Review of Korea*, 1948, Statistics, III, 27-28. Column (6); See note 23, Chapter II.

Table A-11
PRICE INDICES
(base year 1936=100)

Year	General[a] (1)	Agricultural Product (2)	Manufactured Product (3)
1910	54	36	60
1911	59	47	61
1912	62	59	64
1913	64	61	63
1914	60	49	60
1915	56	43	61
1916	65	51	78
1917	92	71	90
1918	123	101	124
1919	155	157	150
1920	161	138	201
1921	121	97	157
1922	121	110	141
1923	119	108	134
1924	128	130	141
1925	136	124	138
1926	123	114	121
1927	115	104	114
1928	113	107	114
1929	109	97	106
1930	95	57	96
1931	77	64	85
1932	76	72	88
1933	84	74	96
1934	85	91	94
1935	94	95	101
1936	100	100	100
1937	121	100	125
1938	139	113	156
1939	164	141	175
1940	180	169	191
1941	187	163	197
1942	196	168	209
1943	215	172	225
1944	241	186	251

Sources: Column (1), *Annual Economic Review of Korea*, III, 426-35.
Column (2) - (3), See Chapter II.
a. Seoul consumer price index.

Table A-12
NET VALUE OF COMMODITY PRODUCT
AT 1936 CONSTANT PRICES
(in million K¥)

Year	Agriculture	Forestry	Fishery	Sub-Total (1)+(2)+(3)
	(1)	(2)	(3)	(4)
1910	553	48	15	616
1911	633	38	13	684
1912	616	31	13	660
1913	697	33	16	746
1914	764	42	22	828
1915	775	48	28	851
1916	805	43	28	876
1917	798	33	27	858
1918	884	25	29	938
1919	688	17	26	731
1920	841	20	26	887
1921	872	53	42	967
1922	845	60	39	944
1923	844	64	44	952
1924	790	52	37	879
1925	853	39	39	931
1926	866	47	45	958
1927	925	56	58	1,039
1928	805	55	60	920
1929	850	69	65	984
1930	1,062	100	83	1,245
1931	947	84	69	1,100
1932	1,022	69	60	1,151
1933	1,056	115	68	1,239
1934	961	105	64	1,130
1935	1,029	108	74	1,211
1936	907	106	87	1,100
1937	1,141	116	91	1,348
1938	1,059	125	87	1,271
1939	756	123	120	999
1940	903	126	134	1,163

Table A-12 continued

Mining	Manufacturing	Sub-Total (5)+(6)	Grand Total (4)+(7)
(5)	(6)	(7)	(8)
8	21[a]	29	645
8	21	29	713
9	17	26	686
10	25	35	781
11	25	36	864
14	38	52	903
14	38	52	928
15	48	63	921
20	56	76	1,014
14	69	83	814
10	44	54	941
8	59	67	1,034
8	66	74	1,018
10	76	86	1,038
11	80	91	970
12	91	103	1,034
16	112	128	1,086
17	120	137	1,176
19	124	143	1,063
20	137	157	1,141
21	121	142	1,387
20	123	143	1,243
31	143	174	1,325
40	152	192	1,431
59	185	244	1,374
70	245	315	1,526
88	290	378	1,478
99	298	397	1,745
119	275	394	1,665
105	319	424	1,423
134	364	498	1,661

Sources: Tables A-3, A-7, A-8, A-9, A-11.

a. Assumed to be the same as 1911 value.

Appendix B

Table B-1

FLOW OF NON-MANUFACTURED DOMESTIC GOODS TO CONSUMERS
(in million K¥ at current prices)

	Agriculture (1)	Forestry (2)	Fishery (3)
1919	967.3	26.6	54.0
1920	1,111.3	27.6	47.2
1921	739.3	51.8	54.7
1924	857.2	70.6	64.6
1925	903.4	45.4	65.0
1926	846.7	53.0	68.2
1929	623.5	63.0	83.7
1930	468.7	54.4	63.1
1931	463.7	51.0	59.1
1934	661.3	70.6	76.6
1935	745.8	76.5	91.6
1936	789.1	76.2	112.4
1939	995.7	144.8	218.2
1940	1,308.2	97.7	254.6

Notes by column:

(1) The entries represent market values of rice, barley, wheat, naked barley, soybeans, red beans, other pulses, Italian millet, barnyard millet, glutenous millet, other cereals, vegetables, fruits. Domestic production minus "flow to food industry." For sources, see notes to Table A-1.

(2) The entries represent market values of forestry products used as fuel and medicinal herbs. For sources, see notes to Table A-1.

(3) Marine products, cultured products, and processed products. Domestic production minus "flows to processed products." For sources, see Table A-8.

Table B-2
DOMESTIC PRODUCTION
OF MANUFACTURED CONSUMERS' GOODS, 1919-1940
(in million K¥, market values at current prices)

	1919 (1)	1920 (2)	1921 (3)	1924 (4)	1925 (5)	1926 (6)	1929 (7)
Textile Products	75.8	56.2	55.4	56.4	61.5	66.2	52.6
Clothing	34.0	20.5	22.7	24.1	27.2	31.9	30.4
Thread, Knit, Cloth & Goods, Others[a]	41.8	35.7	32.7	32.3	34.3	34.3	22.2
Ceramics[b]	8.0	11.1	11.0	10.8	11.7	12.9	6.9
Chemical Products	14.8	10.7	9.3	8.5	15.8	13.5	14.7
Pharma-ceutical Products	4.9	2.2	1.9	3.2	8.3	4.2	6.4
Soap & Cosmetics	0.4	0.5	0.4	0.7	1.2	1.3	3.6
Papers	8.7	7.4	6.7	4.2	5.8	7.6	4.5
Candles	0.8	0.6	0.3	0.4	0.5	0.4	0.2
Wood Products	5.2	5.3	5.8	7.3	7.7	8.3	8.6
Foods	75.3	67.6	70.2	100.2	93.6	87.6	115.5
Others	22.7	26.5	19.7	29.0	31.7	53.4	61.9
Misc. Products	22.7	26.5	19.7	29.0	31.7	53.4	61.9
Printing & Publishing[c]	-	-	-	-	-	-	-
Total	201.8	177.4	171.4	212.2	222.0	241.9	260.2

Sources by column: (1)-(6) Calculated from Chōsen Sōtokufu, *Chōsen no shōkogyō*, Appendix Table 9.; (7)-(14) All entries are calculated from *Chōsen tōkei nempō:* 1929, pp. 222-237; 1931, pp. 214-237, 814-841; 1934, pp. 166-195; 1935, pp. 166-197; 1936, pp. 160-195; 1939, pp. 126-159; 1940, pp. 126-161.

Table B-2 continued

1930	1931	1934	1935	1936	1939	1940
(8)	(9)	(10)	(11)	(12)	(13)	(14)
40.5	29.1	58.8	66.7	75.6	151.8	184.5
22.1	15.0	35.3	43.8	50.0	102.0	126.3
16.6	12.1	17.3	18.3	19.7	37.2	42.0
1.3	1.6	3.4	4.0	4.8	11.0	12.4
0.5	0.4	2.8	0.6	1.1	1.6	3.8
3.8	3.6	6.5	7.7	10.0	19.8	26.5
11.3	10.1	10.8	13.4	14.4	38.0	50.0
5.5	5.1	4.2	4.8	4.6	13.0	13.7
1.0	0.8	1.0	1.2	2.2	13.5	23.2
4.7	4.1	5.5	7.2	7.4	10.2	11.3
0.1	0.1	0.1	0.2	0.2	1.3	1.8
5.5	5.8	7.3	8.2	9.9	35.0	21.0
82.2	81.0	137.7	169.4	199.9	328.4	373.4
68.4	47.9	84.8	97.8	110.7	173.6	200.7
63.8	43.2	78.8	91.0	103.8	163.3	190.0
4.6	4.7	6.0	6.8	6.9	10.3	10.7
211.7	177.5	305.9	363.2	420.5	746.6	856.1

a. Other textile products excluding all types of yarns. The entries for 1919 through 1926 are total values of the three groups: thread, knit cloth and goods, and other textile products.
b. Market values of ceramic products in the official data except cement, lime, flat glass.
c. No figures are available for 1919 through 1929.

Table B-3

DOMESTIC PRODUCTION
OF MANUFACTURED PRODUCERS' GOODS, 1919–1940
(in million K¥, market values at current prices)

	1919	1920	1921	1924	1925	1926	1929
Metal Products[a]	[13.5	[12.0	[13.5	[13.5	[14.4	[14.5	[13.6
Cast Metal Prod.							
Others							
Machine & Tools	4.7	6.1	6.2	6.0	6.4	6.6	11.4
Chemical Products	0.1	0.3	0.4	3.3	8.2	6.1	10.0
Rubber Products	–	–	0.2	2.0	2.5	3.2	6.3
Paints	–	0.1	–	–	–	–	0.1
Chemicals	0.1	0.2	0.2	1.3	5.7	2.9	3.6
Ceramics	0.1	–	0.5	0.7	0.6	0.6	6.6
Cement	[0.1	[–	[0.5	[0.7	[0.6	[0.6	[6.3
Lime & Glass							0.3
Total	18.4	18.4	20.6	23.5	29.6	27.8	41.6

Source: See the source of Table B-2.

Table B-3 continued

1930	1931	1934	1935	1936	1939	1940
1.2	5.4	8.7	11.8	12.3	26.8	28.2
1.2	1.3	2.5	2.9	3.0	8.2	12.5
–	4.1	6.2	8.9	9.3	18.6	15.7
10.1	7.9	9.5	11.5	13.5	53.2	76.7
11.0	18.2	16.9	20.4	31.3	102.8	97.4
4.4	4.6	8.2	10.7	12.7	18.0	25.8
0.1	–	0.1	0.1	0.2	0.1	0.2
6.5	13.6	8.6	9.6	18.4	84.7	71.4
6.6	5.5	6.0	9.8	11.9	23.5	35.1
5.5	5.3	5.5	9.5	11.3	22.4	33.0
1.1	0.2	0.5	0.3	0.6	1.1	2.1
28.9	37.0	41.1	53.5	69.0	206.3	237.4

a. Data for 1919- 1926 include all the metal products.

Table B-4

EXPORTS OF CONSUMERS' GOODS, 1919–1940

(in million K¥, market values at current prices)

		1919	1920	1921	1924	1925	1926	1929
1.	Agricultural Prod. (Crude Food Stuffs)	136.5	98.7	114.3	194.2	199.7	220.0	175.4
2.	Fishery Products	16.4	17.2	18.6	24.2	23.2	25.1	26.8
3.	Manufactured Products	4.5	5.8	23.5	22.1	39.9	38.8	48.8
3.1	Textile Products	2.1	2.5	14.7	10.0	28.4	29.6	32.5
3.2	Food Products	2.2	1.7	1.6	8.0	6.4	3.8	2.4
3.3	Chemical Products (Soap & Paper)	0.2	0.2	0.2	0.1	0.2	0.1	0.6
3.4	Ceramics (China & Clay Prod.)	–	–	–	0.6	0.7	0.4	0.1
	Others	–	1.4	7.0	3.4	4.2	4.9	13.2
4.	Total	157.4	121.7	156.4 ↑ (1, 2, 3 only)	240.5	262.8	283.9	251.0

Table B-4 Continued

1930	1931	1934	1935	1936	1939	1940
140.1	150.0	249.2	268.3	286.3	221.1	60.1
20.6	18.7	27.4	29.4	30.3	65.2	88.2
35.7	32.7	46.0	45.3	55.4	93.9	106.1
26.6	21.6	20.6	22.7	24.4	38.1	42.7
3.0	2.2	5.1	6.3	9.1	8.7	8.1
0.2	-	4.6	3.9	4.5	13.1	14.7
0.4	0.6	1.0	1.0	0.9	1.3	1.2
5.5	8.3	14.7	11.4	16.5	32.7	39.4
196.4	201.4	322.6	343.0	372.0	382.2	254.4

Sources, by line: *Chōsen no shōkogyō,* from which all entries are derived for 1919, Appendix Table 5.
1, 2, 3.2 and 3.5—Derived from the Bank of Korea, *The Annual Economic Review of Korea*, 1948, Statistics, III, 48.
3.1, 3.3 and 3.4—Derived from *Chōsen tōkei nempō*, The Listing of Major Export Products: 1926, pp. 272-277; 1931, pp. 291-293; 1936, pp. 225-228; 1940, pp. 187-190.

Table B-5
IMPORTS OF CONSUMERS' GOODS, 1919-1940
(in million K¥, market values at current prices)

		1919	1920	1921	1924	1925	1926	1929
1.	Agricultural Prod. (Crude Food Stuffs)	21.5	32.2	7.3	45.3	63.6	64.4	24.6
2.	Fishery Products	3.3	2.7	2.4	5.7	5.0	5.6	4.9
3.	Manufactured Products	87.3	71.4	94.1	133.3	150.4	155.8	159.6
3.1	Textile Products	70.6	48.7	59.9	92.7	108.1	110.1	96.4
3.2	Food Products	11.1	12.5	14.4	22.5	23.9	24.3	33.3
3.3	Chemical Products (Soap & Paper)	4.1	5.1	5.8	7.3	7.6	8.2	9.1
3.4	Ceramics (China & Clay Products)	1.5	1.4	1.3	1.8	2.0	2.4	2.9
	Others (Misc. Products)	-	3.7	12.7	9.0	8.8	10.8	17.9
4.	Total	112.1	106.3	103.8	184.3	219.0 (1, 2, 3 only)	225.8	189.1

Table B-5 continued

1930	1931	1934	1935	1936	1939	1940
38.4	18.4	35.6	63.4	63.6	75.0	84.2
4.1	3.7	6.7	6.7	8.0	12.5	9.8
141.2	109.3	170.8	172.9	173.4	368.3	401.5
87.7	67.1	104.1	103.9	102.2	186.9	202.2
27.6	22.0	25.3	41.4	40.5	66.9	76.2
8.2	6.9	12.4	14.9	16.8	28.4	30.1
2.3	2.1	3.9	5.3	5.7	12.2	14.8
15.4	11.2	25.1	7.4	8.2	73.9	78.2
183.7	131.4	213.1	243.0	245.0	455.8	495.5

Sources by Line: All entries for 1919 are derived from *Chōsen no shōkogyō,* Appendix Table 6.
1, 2, 3.2 and 3.5—Calculated from *Annual Economic Review of Korea*, 1948, Statistics, III, 49.
3.1, 3.3 and 3.4—Calculated from *Chōsen tōkei nempō*, 1926, 1931, 1936, 1940.

Table B-6

EXPORTS AND IMPORTS OF PRODUCERS' GOODS, 1919-1940

(in million K¥, market values at current prices)

	1919	1920	1921	1924	1925	1926	1929
Exports	-	-	1.1	1.7	1.4	1.6	4.0
1. Machines & Tools	-	-	0.2	1.2	1.0	1.1	2.0
2. Cement	-	-	0.9	0.5	0.4	0.5	2.0
Imports	18.2	22.4	32.5	31.1	27.8	42.4	63.4
3. Machines & Tools	12.5	13.8	14.0	17.6	14.4	19.6	28.6
4. Metal Products	4.1	6.3	15.6	10.9	11.0	19.3	30.2
5. Ceramics (Cement & Plate Glass)	1.6	2.2	2.6	2.2	2.0	3.0	3.9
6. Chemicals	-	0.1	0.3	0.4	0.4	0.5	0.7

Table B-6 continued

1930	1931	1934	1935	1936	1939	1940
3.5	2.9	4.2	6.4	8.4	6.0	1.3
1.7	1.6	2.4	2.7	2.5	0.9	1.2
1.8	1.3	1.8	3.7	5.9	5.1	0.1
61.6	44.1	95.2	94.1	117.5	251.4	404.0
30.9	21.3	43.2	56.5	68.4	107.4	259.7
26.6	19.8	45.5	31.8	41.0	136.2	138.7
3.2	2.2	5.5	4.9	7.4	5.8	3.6
0.9	0.8	1.0	0.9	0.7	2.0	2.0

Source, by Line: All entries for 1919 are derived from *Chōsen no shōkogyō*, Appendix Table 5 and 6.
1, 2, 5, and 6—Calculated from *Chōsen tōkei nempō*, 1926, 1931, 1936, 1940.
3 and 4—*Annual Economic Review of Korea*, 1948, Statistics, III, 49.

Note: Type of producers' goods is limited to the official listing of "Major Export and Import Products" in *Chōsen tōkei nempō*.

NOTES

Chapter I

1. "Transition" here refers to a political change, which must be distinguished from an economic "transition" to modern economic growth as defined below:

   > Is it really necessary to move back farther and farther in time in order to appreciate the significant dimensions of a recent and entirely different past? We do not believe so, although we do believe that it is important to determine just how far back it is useful to trace the antededents of modern economic growth. This function, we hope, can be performed by the concept of 'transition.'
   >
   > Henry Rosovsky, "Japan's Transition to Modern Economic Growth," In Henry Rosovsky, ed., *Industrialization in Two Systems,* Essays in Honor of Alexander Gerschenkron, (New York, 1966), p. 92.

2. Scholars of Korean economic history generally take the year 1876 as the beginning of the modern period. For further discussion of classifying periods in Korean economic history, see Ki-jun Cho, *Hanguk kyūngje sa* (Seoul, 1954), pp. 30–37.

3. As a general reference on Korean history for the period covered here, see C.N. Weems, ed., *Hulbert's History of Korea* (New York, 1962) Vol. I, and Vol. II.

4. They may be contrasted with the favorable effects of the isolation period in Japan. See Henry Rosovsky, *Capital Formation in Japan, 1868–1940* (Glencoe, 1961), pp. 85–88.

5. Professor Alexander Gerschenkron points out that the realization of tension between the actual state of economic conditions and the great promise of modern economic growth played a vital role in the development of many nations. See his "Economic Backwardness in Historical Perspective," in Bert F. Hoselitz, ed., *The Progress of Underdeveloped Areas* (Chicago, 1952).

6. For the details on the history of land-tenure system in Korea, see Seizo Katō, *Kankoku nōgyō-ron* (Tokyo, 1904).

7. Andrew J. Grajdanzev, *Modern Korea* (New York, 1944), p. 25.

8. For details on the *Donghark* movement, see Ki Baek Lee, *Hanguk sa sin lon* (Seoul, 1969), pp. 314–320.

9. Hilary Conroy, *The Japanese Seizure of Korea: 1868–1910* (Philadelphia, 1960), p. 459.

10. See Hiroshi Shikata, "Chōsen ni okeru kindai shihonshugi no seiritsu katei" in *Chōsen shakai keizai-shi kenkyū* (Tokyo, 1933), pp. 171–178.

11. Ibid., p. 168.

12. Conroy, p. 448.

13. 1 *chungbo* = 3.95 square yards.

14. Zenkoku Keizai Chōsa Kikan Rengōkai, Chōsen Shibu, *Chosen keizai nempō, 1939* (Tokyo, 1939), p. 39.

15. Conroy, p. 472.

16. For a brief discussion on the origin of banks and the currency system in Korea, see the Bank of Chōsen, *Economic History of Chōsen* (Seoul, 1920), pp. 43–72.

17. Kurō Kobayakawa, ed., *Chōsen nōgyō hattatsu-shī* (Seoul, 1944), p. 353.

18. Conroy, p. 450.

19. Ibid., pp. 477–479.

20. Toyomasa Yamaguchi, *Chōsen no kenkyū* (Tokyo, 1911), p. 228.

21. Hilary Conroy argues that "economic matters had no important effect in determining the Japanese course toward annexation of Korea," p. 491.

22. The characteristics of the Government-General of Korea are well summarized in the following: "Theoretically, the post of Government-General is open to Japanese civilians as well as to generals and admirals, but during the thirty-two years that have passed since the annexation of

Korea, no civilians had ever held this office. But whatever this rank and power may be in Tokyo, in Korea the Government-General is virtually an absolute monarch; he is the head of the administration including the police; he is also the law maker." Grajdanzev, p. 238.

23. How the colonial period is divided depends on the writer's viewpoint. For instance, Suzuki divides the period into the following intervals: 1910–1920 for basic foundation of the economy; 1920–1930 for developing primary industry including mining for export purposes; 1930–1936 for industrialization; and 1937–1945 for war preparation. See Takeo Suzuki, *Chōsen no keizai* (Tokyo, 1942), pp. 25–28.

24. K. Ohkawa and H. Rosovsky "A Century of Japanese Economic Growth," in W. W. Lockwood, ed., *The State and Economic Enterprise in Japan* (Princeton, 1965), p. 75.

25. William Lockwood, *The Economic Development of Japan: Growth and Structural Changes, 1868–1938* (Princeton, 1954), p. 18.

26. See Ohkawa and Rosovsky, "The Role of Agriculture in Modern Japanese Economic Development," *Economic Development and Cultural Change,* IX.I (October 1960).

27. Toyomasa Yamaguchi estimated that the cultivated area of Korea during 1907 could be increased by 67% by simple reclamation of wasteland. See Yamaguchi, pp. 105–06.

28. John K. Fairbank, Edwin O. Reischauer, Albert M. Craig, *East Asia: The Modern Transformation* (Boston, 1965), p. 760.

29. The major effects of the civil law on the rural economy are discussed in Kuranozō Tsumagari, "Chōsen ni okeru kosaku mondai no hatten katei," *Chōsen keizai no kenkyū* (Tokyo, 1929), pp. 325–385.

30. The Japanese farmers among the landlord class acted as leaders in adopting advanced techniques of cultivation from Japanese agriculture. Indeed, the Japanese farmers residing in Korea had dual roles: borrowing advanced techniques from Japan, and controlling the rural inhabitants in accordance with the policy objectives of the colonial administration.

31. These figures are taken from the Statistical Yearbook of the Government-General.

32. For details on the role and achievement of these institutions, see Hoon K. Lee, *Land Utilization and Rural Economy in Korea* (Hong Kong, 1936), pp. 281–289.

33. John C. H. Fei and Gustav Ranis, *Development of the Labor Surplus Economy* (Homewood, 1964), pp. 125–131.

34. Ohkawa and Rosovsky, "Century of Growth," p. 79.

35. See Fairbank, Reischauer, and Craig, pp. 568–579.

36. Ohkawa and Rosovsky, "Century of Growth," p. 81.

37. William W. Lockwood, *The Economic Development of Japan* (Princeton, 1965), pp. 64–70.

38. See E. B. Schumpeter, ed., *The Industrialization of Japan and Manchuria* (New York, 1940), pp. 741–786.

39. Ibid., p. 371.

Chapter II

1. When countries are grouped by per capita income level, low income countries tend to have a relatively small 'S' sector in the distribution of national product and labor force. See Simon Kuznets, *Six Lectures on Economic Growth* (Glencoe, 1960), Table 5, pp. 45–47.

2. For example, Shannon McCune praises highly the statistical data compiled by the Government-General of Chosun. See his *Korea's Heritage: A Regional and Social Geography* (Tokyo, 1956), Appendix on bibliography.

3. The original and revised data for rice production are:

	Paddy Field (1,000 *changks*)			Volume of Rice Production (1,000 *suk*)		
Year	Original	Revised	% Increase	Original	Revised	% Increase
1910	826	1,353	64	7,918	10,406	13
1911	962	1,399	45	9,973	11,568	12
1912	982	1,417	44	8,982	10,865	12
1913	1,046	1,457	39	10,091	12,109	12

1914	1,079	1,484	38	12,159	14,131	12
1915	1,128	1,498	33	11,374	12,846	11
1916	1,158	1,519	31	12,531	13,933	11
1917	1,169	1,529	31	12,227	13,688	11
1918*		1,548			15,294	
1919		1,538			12,708	

*The year when the land survey was completed.

4. The Korean case was quite different from the Japanese experience of underreporting in agriculture during the late 19th century (as pointed out by Professor James Nakaman in his *Agricultural Production and Economic Development of Japan*).

5. See H. Lee, pp. 60–61.

6. See Chōji Hishimoto, *Chōsen-mai no kenkyū* (Tokyo, 1940), pp. 673–674.

7. For the details concerning the new survey, see Seichi Tohate and Sasushi Ohkawa, *Chōsen beikoku keizairon* (Tokyo, 1939), pp. 425–29.

8. Ideally, the cost of "buildings" in Table 2 should be included in our calculations of the cost of rice production. However, this item seems to be unusually high in Table A-3 (cf. y .44 in Japan), and it shows a wide variation during the subsequent years. In fact, the official imputation for the cost of "buildings" is the least accurate data, often used to demonstrate the high-cost aspect of Korean rice in order to reduce Japan's restrictions on imports of Korean rice during the early 1930s. For this reason, we omitted this item from the calculation of cost per unit of rice production.

9. The average price in the nine large cities in Korea.

10. It should be noted that the cost of fertilizer accounted for about 70% of the cost of rice productions in 1933.

11. Professor Ohkawa used an upper limit of 88% for Japanese agriculture of 1878. In selecting an upper limit of 90% for Korean agriculture, our assumption is that Korean agriculture in the 1910s was less developed than Japanese agriculture of 1878, in the sense that a larger portion of inputs was provided by the farmers themselves.

12. The index of fertilizer covers both the commerical fertilizer and the manure produced by the farmers themselves. The following weights among the different types of fertilizers are used, based on their market values during 1933–1937: compost, 61%; commercial fertilizer, 33%; and green manure, 6%.

13. The definition was modified in 1929 to exclude metal refineries and lumbering establishments which had less than 30 employees. The output of government-owned factories was excluded from the official data, except during 1923–1928.

14. A closer observation of the products listed under "manufactured commodities" reveals that the following items of factory output were omitted from the list until 1919: metal refinery, manufactured salt, printing, gas and electricity, dyeing and cloth-making. Thus, the market value of these items is also subtracted from the factory-output data: rice processing, lumbering, and cotton processing.

15. Rice processing in Japan is excluded from the list of factory output, whereas it constitutes the major item in the Korean list in food and beverages.

16. Kazushi Ohkawa et al., *The Growth Rate of the Japanese Economy Since 1878* (Tokyo, 1957), p. 87.

17. Ibid., p. 90.

18. Yamada's estimates of net income ratios for marine products ranged from 69% to 64% during 1903–1948, with a single ratio of 35% for the processed products, and 40% for salt. *Nihon kokumin shotoku suikei shiryō* (Tokyo, 1937).

19. Ibid., pp. 42–43.

20. We omit here a discussion on the major bias inherent in the long-run estimates of national product in general. See Simon Kuznets, ed., *Income and Wealth of the United States,* Series II, (Baltimore, 1952), pp. 40–47.

21. The over-all index was constructed by taking the simple arithmetic average of 30 selected commodities. See the Bank of Korea, *The Annual Economic Review of Korea,* 1949, pp. 425–35.

22. See Chōsen Ginkō (Bank of Korea), *Tokai geppō* (Bulletin of monthly statistics) 1933–1941.

23. To derive a single index from the 5 indices, the relative weights were changed every 5 years starting from 1910. Within each 5-year interval, the 5-year averages of current values are used to determine the relative weights.

24. They are cotton yarn, drapery (Japanese), oil, sugar, soy sauce, (table) salt, sake (rice wine), beer, printing paper, and matches.

25. The relative weights were changed every 5 years, starting from 1911–1915. Within each 5-year interval, the 5-year average of current values is used to determine the relative weights.

Chapter III

1. A "city" was an officially designated administrative unit, based on the size of population.

2. According to Professor Rosovsky, 1868–1885 saw Japan's transition to modern economic growth. See his article "Japan's Transition to Modern Economic Growth," in H. Rosovsky, ed., *Industrialization in Two Systems* (New York, 1966), p. 92.

3. The figures for the Korean economy are taken from Table A-12.

4. Rosovsky, "Japan's Transition," p. 93.

5. The census was taken in 1944, instead of 1945, because of Japan's worsening position in World War II.

6. Grajdanzev points out that "the authorities seem to adjust their current statistical figures (of population) to those of the census, though they still lag behind," p. 72.

7. The official estimate of Korean population for 1910 was 13,129,000, compared with 15,958,000 for 1915. This shows a net increase of 21.5% during the 5-year period, which seems to be unusually high compared to the growth rates of the subsequent years for which more accurate data are available. It is evident that such a high rate of growth was due largely to an underestimate in 1910, when the colonial

administration was in an early stage. During 1925–1944 the average percentage growth of population was about 15% per decade. If one extrapolates backward from the 1930s census data, this rate yields a 1910 population of 14,766,000.

8. Until the 1930s, the movement of Koreans across national boundaries was negligible.

9. The figures in this table include both Korean and Japanese population in Korea. Therefore, the rate of actual increase does not coincide with Table 13.

10. General explanations for these trends may be found in Japan's experience of population growth during her early years of modern industrialization. "Order, economic development, and medical technologies permitted the decline of death rates, while ancient ways of living and thinking among the peasants insured that birth rates remained at or near the levels that had been essential to biological and cultural survival in earlier centuries." Irene B. Taeuber, "Population and Labor Force in the Industrializing Japan, 1850–1950" in *Economic Growth: Brazil, India, Japan* (Durham, N.C., 1955), p. 316

11. One of the policy measures taken for the rapid expansion of population was to make annual rewards to those parents who had more than a certain number of children.

12. The United Nations, *Population Bulletin, No. 1,* December 1951.

13. By the term "urban area" we mean the regions designated as *fu* (city) by the government. There were 18 such cities in 1939.

Chapter IV

1. The adjustment "for price changes of the product requires not the usually available prices of the commodities and services produced in industry but prices of that part of the product which constitutes the net income." National Bureau of Economic Research, Bulletin, No. 59 (May 1936), quoted by Simon Kuznets, "Long-Term Changes in the National Income," p. 93.

2. The contribution rate of a given sector toward growth of over-all totals is calculated by $1 + (P_b/P_a \times r_b/r_a)$ where P_a=product of a given sector,

P_b=product of all other sectors, r_a=growth rate of P_a, r_b=growth rate of P_b.

3. Ohkawa, et al., *Growth Rate of the Japanese Economy,* p. 142.

4. For a brief history of population census during the colonial period, see Economic Planning Board, Republic of Korea, *Statistical Monthly,* Nos. 1-2 (1962), pp. 26-47.

5. "Primary worker" refers to a full-time male worker and "secondary worker" to a female worker or a minor.

6. The crude activity rate is defined as "the percentage of the economically active population to the total population of all ages." The United Nations, Department of Economic and Social Affairs, *Demographic Aspects of Manpower*, Report I, (New York, 1962), p. 3.

7. In the official data, the Japanese population in Korea was classified by the type of occupation of the householder.

8. The distribution of Japanese in the agricultural sector during the early years of the colonial period should be considered relatively high in view of their special status in the country (cf. industrial distribution of Japanese workers in Formosa during the colonial period). See Barclay, p. 67.

9. During 1930-1940, the Japanese population in Korea showed a 34.2% increase, whereas the Korean population increased by 15.2%.

10. Simon Kuznets, *Commodity Flow and Capital Formation,* Vol. I (New York, 1938).

11. In the present analysis, Professor Simon Kuznets's definitions are adopted for finished and unfinished goods, consumers' and producers' goods. Ibid., p. 6.

12. Mixed commodities may be defined as commodities of which a portion goes to ultimate consumers and the rest to producers as unfinished (or semi-finished) goods.

13. In the absence of relevant data for allocation of mixed commodities among different uses, we had to classify them as representing one type

of commodities (i.e. a finished commodity or unfinished commodity, consumers' goods or producers' goods).

14. The portion of crude stuffs used by producers for manufacturing of food products was roughly estimated by the differences between manufactured foods at producer prices and their net values in Chapter II. These differences were subtracted from the total values of crude stuffs in order to estimate finished goods in agriculture.

15. The livestock products are a typical case of mixed commodities. Some of the commodities (i.e. meat, eggs) go to ultimate consumers. However, a large portion of them are used for manufacturing foods, various leather products, etc. Thus, we decided to classify livestock products as unfinished goods.

16. In Chapter II, the net product ratio of processed products in fishery was 35% of their values at producer prices. Therefore, 65% of the processed products at producer prices were subtracted from marine products in estimating finished goods.

17. Since the present analysis does not include construction, all the building materials are classified as finished goods.

18. All types of oils and gasoline (including those used for illuminating) should be classified as mixed commodities. However, we decided to treat them as unfinished goods for the reasons indicated in Note 13 above.

19. The market value of printing and publishing included the value of papers used in this sector. Therefore, we decided to estimate the value of finished product in this sector by its net value added.

20. The main products belonging to this category were cigarettes (manufactured), leather products, straw products, etc.

21. For a survey of income elasticities of demand for food at different income levels, see S. Kuznets, "Quantitative Aspects of the Economic Growth of Nations: The Share and Structure of Consumption," *Economic Development and Cultural Change,* X.2, Part 2 (January 1962), p. 25.

22. In view of the fact that, during the early years, the ratio of foodstuffs to clothing was extremely high (see Table 29), it seems evident that

a substantial amount of handicraft products (e.g. clothing) was not recorded in the statistical data.

23. One of these organizations was Chōsen Kin'yū Kumiai (Financial Cooperative of Korea), which dealt exclusively with farmers on matters of finance, purchase, and sale of goods.

24. Some of the main products replaced by the marketed (manufactured) goods were beverages, clothing, shoes, furniture, etc.

25. See Kuznets, "Quantitative Aspects," pp. 36–48.

26. Ibid., p. 3.

27. See Table 63.

28. For this purpose, our analysis is carried out on a selected-year basis, determined by the availability of data.

Chapter V

1. Rosovsky, *Capital Formation in Japan,* p. 83.

2. For details of the agricultural policy, see Kurō Kobayakawa, ed., *Chōsen nōgyō hattatsu-shi*, Vol. II, *Policy*.

3. The Oriental Development Company was established as a joint-stock company during 1908–1909 by the Oriental Associations in Japan for the purpose of expanding cultivated areas and developing agriculture in Korea. However, it had been the major instrument in expanding Japanese ownership of land and controlling the agricultural sector in accordance with government policies. Its operation was heavily subsidized by the Japanese Government. For the official statement on the purpose of the company, see the Government-General of Chōsen, *Annual Report on Reforms and Progress in Korea,* 1908–1909, pp. 16–20.

4. The official publications (e.g., *Chōsen tokei nempō*) do not show the amount of arable land by nationality or size of holding. Some aspects of the Japanese ownership of land in Korea are discussed in Chong Sik In, *Chōsen no nōgyō kikō* (Tokyo, 1940), pp. 289-298.

5. Bruce F. Johnston, *Japanese Food Management in World War II* (Stanford, 1953), p. 55.

6. The Government-General also adopted Japan's civil law in 1912. This reaffirmed the private ownership of land, including Japanese, and provided unlimited guarantees for private property.

7. Through the land survey, most of the arable land which belonged directly to the Korean monarchy prior to the annexation was sold to private individuals (mainly Japanese).

8. For details on tenancy system in Korea during the colonial period, see Chōsen Sōtokufu, *Chosen no kosaku kankō* (Seoul, 1930).

9. Hoon K. Lee, p. 163.

10. Kamekichi Takahashi, *Gendai Chōsen keizairon* (Tokyo, 1955), pp. 189–193.

11. Seibun Suzuki, *Chōsen keizai no gendankai* (Seoul, 1930), pp. 456–468.

12. Hoon K. Lee, *Land Utilization,* p. 137.

13. See Gustav Ranis, "The Financing of Japanese Economic Development," *Economic History Review* XI.3 (April 1959).

14. *Economic Growth and Structure,* p. 250.

15. The income elasticity for food of 0.7 was used by Ohkawa and Rosovsky in their estimates of Japan's demand for food. See their article, "The Role of Agriculture in Modern Japanese Economic Development," *Economic Development and Cultural Change,* IX.I, Part 2 (October 1960).

16. The conversion of food grains into calories used the conversion ratio suggested by U.S. Department of Agriculture Handbook No. 34 (1952).

17. It may very well be that the share is negatively related to the per capita income level: the share may go up as per capita income deteriorates.

18. See pp. 79–83.

19. Johnston, *Japanese Food Management,* p. 55.

20. Chong Sik In, p. 331.

21. Johnston, *Japanese Food Management,* p. 54.

22. Pp. 17–19.

23. Various measures were taken to reduce domestic consumption to the minimum and to increase commodity reserves for the preparation for war. See *Chōsen keizai nempō,* 1939, pp. 420–433.

24. Johnston, *Japanese Food Management,* p. 55.

Chapter VI

1. In contrast to the Korean experience, it is well known that the central government of Japan during its early years of modern economic growth was instrumental in channeling agricultural surplus into financing industrial development. See Ranis, "The Financing of Japanese Economic Development."

2. See Naoji Kada, *Chōsen kōgyō kihon chōsa gaiyō* (Seoul, 1934).

3. See Akitake Kawai, *Chōsen kōgyō no gendankai* (Seoul, 1943), pp. 96–111.

4. Ibid., p. 120.

5. Chōsen Sōtokufu, *Chōsen no shokogyō*, 1929, pp. 25–30.

6. Ohkawa and Rosovsky, "Century of Growth," p. 82.

7. Concerning the business activities and economic consequences of *zaibatsu,* see Lockwood, *The Economic Development of Japan,* pp. 214–235.

8. Grajdanzev, p. 159.

9. See Ohkurasho, Kanrikyoku (Control Bureau, Treasury of Japan) *Nihonjin no kaikai kazudo ni kansuru rekishi deki chosa Chōsen hen, Kōgyō no hatasu*, pp. 17–24.

10. See Chōsen Shokusan Ginkō (Industrial Bank of Korea), "Rinji shikin choseiho to Chōsen keizai," *Shoku gin geppō,* June, July, and August issues of 1938.

Chapter VII

1. H. K. Lee, p. 42.

2. See Table 52.

3. For the origins of industrial dualism in Japan, see Seymour Broadbridge, *Industrial Dualism in Japan* (Chicago, 1966), pp. 8–26.

4. Until the annexation in 1910, trade statistics of Korea were compiled by the Yi dynasty, and issued publically in *Trade Report.*

5. Compared to the normal concept of balance of payments, the data suffer from omissions of services, unilateral transactions, and capital accounts.

6. In computing trade ratios, the GNP of Korea is estimated from net commodity-product of Appendix A.

7. The export figures in Table 58 show a relatively high ratio of finished manufactured product during the early as well as the later years of the period under review. However, it should be noted that the exports of the early years were mainly handicraft products, whereas those of the later years were the products of heavy industries.

8. For the real capital flows of 1910 and 1941, the annual averages of 1911–1915 and 1936–1940 in Table 63 are used.

9. Seoul Chamber of Commerce, *Chōsen ni okeru naiji shihon no toka genkyō* (Seoul, 1944), pp. 39–42.

Chapter VIII

1. Since most of the quantitative data used in the preceding analysis are available by provinces, which belong to either South Korea or North Korea today, they are grouped in terms of the two regions in the present chapter. Roughly, South Korea comprises 7½ provinces, and the remaining 5½ belong to North Korea. They are:

   South Korea–Kyungsang namdo, Kyungsang bookdo, Chunla namdo, Chunla bookdo, Choongchung namdo, Choongchung bookdo, Kyunggi do, and half of Kangwon do.

   North Korea–Hamkyung namdo, Hamkyung bookdo, Pyungan

namdo, Pyungan bookdo, Whanghai do, and half of Kangwon do.

2. For a brief historical account of the partition of Korea, see George H. McCune, *Korea Today,* (Cambridge, Mass., 1950), pp. 52-60.

3. Since emigration of Koreans during the 1930s took place from both regions, it seems reasonable to assume that emigration was not the major factor determining differences in growth rates of population between south and north during the colonial period.

4. The figures in Table 65 as a whole reflect both high birth rates and a declining trend of infant death rates during the later years of the colonial period in south and north Korea. "The political and economic policies of the Imperial Japanese Government and the Government General of Chosen did much to maintain the high rate of population increase (in Korea) that was long deplored and feared by the Japanese." Irene B. Taeuber and George Barclay, "Korea and the Koreans in the Northeast Asian Region," *Population Index,* XVI.4 (October 1950), 282-283.

5. In the official publications available to us (i.e., *Chōsen tokei nempō*), agricultural and forestry products are not classified by province until 1934.

6. In Table 68, the relative shares of manufactured product in the two regions show drastic changes during 1934–1935. This was caused by the expansion of chemical products in the north during 1935 through the openings of such new factories as the Korean Coal Corporation, The Korean Nitrogen Gunpowder Operation, the Soya Bean Chemical Industry. See *Chōsen keizai nempō,* 1939, pp. 219–222.

7. According to official data (*Chōsen tōkei nempō*), the south produced about 70% of total "manufactured commodities" prior to the 1930s.

8. See the agricultural commodity products in Table A-12.

9. Producers' goods are here defined to include intermediate goods destined for the use of producers. Thus defined, they include metals, machines and tools, and chemical products (largely fertilizers, chemicals for industrial use, etc.).

10. A striking feature of Table 74 is that chemical products alone accounted for over 61% of total manufactured product in the north during

1939–1940. Some of the major factories in this sector during the 1930s were the Nitrogen Fertilizer Plant, the Korean Magnesium Corporation, the Korean Coal Corporation, the Korean Chemical Corporation, etc. See Akitake Kawai, *Chōsen kōgyō no gendankai*, pp. 120–129, 289–299.

11. For example, the distribution of electric power between the two regions in 1945 was as follows:

	North Korea	South Korea
Total Capacity Percentage	1,262,500 KW (86%)	206,290 KW (14%)
Annual Production Percentage	909,200 KW (92%)	79,500 KW (8%)

Chapter IX

1. Professor Simon Kuznets's formulation of modern economic growth may be found in his *Six Lectures on Economic Growth,* Lecture I and "Reflections on the Economic Growth of Modern Nations," *Economic Growth and Structure, Selected Essays* (New York, 1965).

2. *Economic Growth and Structure,* p. 93.

3. T. S. Ashton, "Economic History and Theory," *Economica* XIII. 5 (May 1946), p. 84.

4. See T. Watanabe, "Economic Aspects of Dualism in the Industrial Development of Japan," *Economic Development and Cultural Change,* April, 1965. Broadbridge, *Industrial Dualism in Japan.*

5. Computed from Ohkawa, *The Growth Rate of the Japanese Economy.*

6. See Chapter VI, pp. 103–105.

7. The Bank of Korea, *Annual Economic Review of Korea,* 1949, Statistics, IV, 426–435.

8. Henry Rosovsky, *Capital Formation in Japan, 1868–1940,* p. 83.

9. Seibun Suzuki, pp. 324–326.

10. See Table 14.

11. In the absence of reliable data on birth rates in the two communities, their contrasting patterns may be seen in terms of the number of children (aged 0–4) per 1,000 women (aged 15–49):

	1925	*1930*	*1935*	*1940*
Koreans	706	696	726	731
Japanese	557	543	529	500

Source: Taeuber and Barclay, "Korea and the Koreans," p. 283.

BIBLIOGRAPHY

Abramovitz, Moses, ed. *Capital Formation and Economic Growth.* Princeton, Princeton University Press, 1955.

Allen, George C. *Japanese Industry: Its Recent Development and Present Condition.* New York, Institute of Pacific Relations, 1940.

——— *A Short Economic History of Modern Japan, 1867–1937.* London, George Allen & Unwin, 1946.

Ashton, T. S. "Economic History and Theory," *Economica* XIII.5 (May 1946).

Bank of Chōsen (Korea). *Economic History of Chōsen.* Seoul, 1920.

Bank of Korea. *Monthly Statistical Review,* 1952–1962.

——— *Economic Statistics Yearbook,* 1960–1964.

Barclay, George W. *Colonial Development and Population in Taiwan.* Princeton, Princeton University Press, 1954.

Bowden, Witt, Michael Karpovich, and Abbott Payson Usher. *An Economic History of Europe Since 1750.* New York, American Book Company, 1937.

Broadbridge, Seymour. *Industrial Dualism in Japan.* Chicago, Aldine Publishing Co., 1966.

Bunee, A. C. "The Future of Korea," *Far Eastern Survey,* Vol. 13, 1944.

Cho, Ki-jun 趙璣濬. *Hanguk kyūngje sa* 韓国経済史 (Economic history of Korea). Seoul, 1954.

Cho, Yong Sam. *Disguised Unemployment in Underdeveloped Areas With Special Reference to South Korean Agriculture.* Berkeley and Los Angeles, University of California Press, 1965.

Choe, Ho-jin 崔虎鎭. *Hyundae Chōsun Kyūngje sa* 現代朝鮮

経済史 (Modern economic history of Korea). Seoul, 1951.

Chōsen Ginkō 朝鮮銀行 (Bank of Korea). *Chōsen Ginkō geppō* 朝鮮銀行月報 (Quarterly report of the Bank of Korea). July 1911.

———. *Sen-man Keizai jūnen-shi* 鮮満経済十年史 (Ten-year economic history of Korea and Manchuria). 1919.

———. *Chōsen ni okeru kōsanhin no jukyū to sono shōrai.* 朝鮮に於 工産品の需給とその將來 (Demand-supply conditions and future prospects for manufactured goods). Seoul, 1937.

Chōsen Keizai Kenkyū-sho 朝鮮経済研究所(Institute of Korean Economy). *Chōsen sōran* 朝鮮総覧 (A guide to Korean statistical materials). Seoul, 1931.

Chōsen Kin'yū Kumiai 朝鮮金融組合 (Association of Financial Cooperatives of Korea). *Chōsen kin'yū kumiai kyōkai-shi* 朝鮮金融組合協会史 (History of the Association of Financial Cooperatives of Korea). Seoul, 1935.

Chōsen Kōgyō Kyōkai 朝鮮工業協会 (Korean Industry Association). *Chōsen no kōgyō to sono shigen* 朝鮮の工業とその資源 (Korean industry and resources). Seoul, 1937.

Chōsen Kōsei Kyōkai 朝鮮厚生協会 (Korean Society of Welfare). *Chōsen ni okeru jinkō ni kansuru sho tōkei* 朝鮮に於 人口に関する諸統計 (Statistics on population in Korea). Seoul, 1945.

Chōsen Shokusan Ginkō 朝鮮殖産銀行 (Industrial Bank of Korea). *Chōsen Shokusan Ginkō nijūnen-shi* 朝鮮殖産銀行二十年史 (Twenty-year history of the Industrial Bank of Korea). Seoul, 1938.

Chōsen Sōtokufu 朝鮮総督府 (Government-General of Korea), 1910–1945. *Chōsen tōkei nempō* 朝鮮統計年報 (Statistical yearbook of Korea). 1910–1942. The 1943 yearbook was published by the Provisional

Government in 1948. *Taishō 14 (jūyo)-nen kan'i kokusei chōsa kekka-hyō* 大正十四年簡易国勢調査結果表 (Table of the results of the 1925 simplified census in Korea). Seoul, 1926.

———. *Chōsen chōsa geppō* 朝鮮調査月報 (Quarterly report of Korean research). 1925.

———. *Chōsen no kei* 朝鮮の契 (The financial corporation in Korea). 1925.

———. *Chōsen no kome* 朝鮮の米 (Korean rice). 1926.

———. *Chōsen no bussan* 朝鮮の物産 (The Korean commodities). 1927.

———. *Chōsen no shijō keizai* 朝鮮の市場経済 (Market economy in Korea). 1929.

———. *Chōsen no shōkōgyō* 朝鮮の商工業 (Korean commerce and manufacturing). 1929.

———. *Chōsen no kosaku kankō* 朝鮮の小作慣行 (Tenant customs in Korea). 1930.

———. *Chōsen no nōgyō* 朝鮮の農業 (Korean agriculture). 1930–1942.

———. *Chōsen no keizai jijō* 朝鮮の経済事情 (Economic conditions in Korea), 1931, 1933, 1934.

———. *Chōsen no zeikan* 朝鮮の税関 (Korean tariffs and customs). 1931.

———. *Shōwa 5 (go)–nen Chōsen kokusei chōsa hōkoku* 昭和五年朝鮮国勢調査報告 (Report of the 1930 census in Korea). Seoul, 1932–1934.

———. *Chōsen no sangyō* 朝鮮の産業 (Korean industries). 1933, 1935.

———. *Chōsen ni okeru beikoku tōsei no keika* 朝鮮に於ける米穀統制の経過 (The control of food supply in Korea). 1934.

———. *Chōsen sembai-shi* 朝鮮専売史 (History of the Korean monopoly administration). 3 volumes. 1936.

———. *Chōsen tetsudō-shi sōshi jidai* 朝鮮鉄道史創始時代 (History of the Korean railways' early years). 1937.

———. *Shōwa 10 (jū)–nen Chōsen kokusei chōsa hōkoku* 昭和十年朝鮮国勢調査報告 (Report of the 1935 census in Korea). Seoul, 1937–1938.

———. *Chōsen denseikō* 朝鮮田制考 (Studies in the Korean land-tenure system). 1940.

———. *Chōsen Shōwa 15 (jūgo)–nen kokusei chōsa kekka yōyaku* 昭和十五年国勢調査結果要約 (Summary of the results of the 1940 census in Korea). Seoul, 1944.

———. *Jinkō chōsa kekka hōkoku* 人口調査結果報告

Cole, Ansley J. and Edgar M. Hoover. *Population Growth and Economic Development in Low-Income Countries: A Case Study of India's Prospects.* Princeton, Princeton University Press, 1958.

Conroy, Hilary. *The Japanese Seizure of Korea: 1868–1910.* Philadelphia, University of Pennsylvania Press, 1960.

Dore, R. P. *Land Reform in Japan.* London, Oxford University Press, 1959.

Doutsch, B. and A. Bekstein. "Population Sovereignty and the Share of Foreign Trade," *Economic Development and Cultural Change* X.4 (July 1962).

Economic Research Center of Korea. *Industrial Structure of Korea.* Seoul, 1962.

Fairbank, John K., Edwin O. Reischauer, and Albert M. Craig. *East Asia: The Modern Transformation.* Boston, Houghton Mifflin Company, 1965.

Fei, John C. H. and Gustav Ranis. *Development of the Labor*

Surplus Economy: Theory and Policy. The Economic Growth Center, Yale University. Homewood, Richard D. Irwin, Inc., 1964.

Gerschenkron, Alexander. "Economic Backwardness in Historical Perspective," in *The Progress of Underdeveloped Areas,* ed. Bert F. Hoselitz. Chicago, University of Chicago Press, 1952.

———. "Reflections on the Concept of 'Prerequisites' of Modern Industrialization," *L'industria,* no. 2 (1957).

Government-General of Chōsen (Korea). *Annual Report on Reforms and Progress in Chōsen.* Title changed from 1924: *Annual Report on the Administration of Chōsen.*

Grajdanzev, Andrew J. *Modern Korea.* New York, Institute of Pacific Relations, 1944.

Hanguk Munwha sa. 韓國文化史 *Hanguk Munwha kyūngje silwhang* 韓國文化經濟實況 (Survey of Korean culture and economic conditions). Seoul, 1957.

Hanguk Sanop Ūnhaeng 韓國產業銀行 (Korea Development Industrial Bank). *Sanop Unhaeng wolbo* 產業銀行月報 *(Monthly Research Bulletin),* 1950–1962.

———. *Hanguk Sanop kyūngje sip-nyon sa* 韓國產業經濟十年史 *(Economic Review of Korea, 1945–1955).* Seoul, 1955.

———. *Hanguk Ui Sanop* 韓國의產業 (Korean industries). Seoul, 1958.

Hanguk Ūnhaeng 韓國銀行 (The Bank of Korea). *Hanguk kyūngje tongge yunbo* 韓國經濟統計年報 *(Economic Statistics Yearbook of Korea).* 1948–1940. Title varies: *Annual Economic Review of Korea,* 1955–1959; *Annual Economic Review,* 1960——; *Economic Statistics Yearbook* (in Korean and English).

———. *Chosa wolbo* 調査月報 *(Monthly Statistical Review)* May 1947–May 1950.
Title varies: *Chōsen Ūnhaeng chosa wolbo* 朝鮮銀行調査月報 March 1951–June 1962; *Hanguk Ūnhaeng chosa wolbo* 韓國銀[...] 調査月報 July 1962––.

———. *Hanguk kukminsoduk chugei-bop* 韓國國民所得推計法 (Methods of estimates for Korean national income). 1954.

———. *San-up chongram* 產業總覽 (Review of Korean industries). 1954.

———. *Hanguk kyūngje* 韓國經濟 1956 (The Korean economy, 1956).

———. *Hanguk suchul san-up* 韓國輸出產業 (Export industries in Korea).

———. *Waekuk muyok tongge* 外國貿易統計 (Foreign trade statistics). 1959.

Higgins, Benjamin. *Economic Development.* New York, W. W. Norton and Co., 1959.

Hisama, Ken'ichi 久間健一 *Chōsen nōgyō no kindaiteki yōsō* 朝鮮農業の近代的樣相 (Modern elements in Korean agriculture). Tokyo, 1935.

Hishimoto, Chōji 菱本長次 Chōsen-mai no kenkyū 朝鮮米の研究 (Study of rice in Korea). Tokyo, 1940.

Hoffman, W. G. *The Growth of Industrial Economics.* Translated from the German by W. G. Henderson and W. H. Chaloner. Manchester, Manchester University Press, 1958.

Hong Sung-yu 洪性囿 *Hanguk kyūngje wa mikuk wōnjō* 韓國經[...] 美國援助 (The Korean economy and United States aid). Seoul, 1962.

Hoselitz, Bert F. "Patterns of Economic Growth," *Canadian Journal of Economics and Political Sciences* XXI.4 (November 1955).

———. "Population Pressure, Industrialization, and Social Mobility," *Population Studies* XI.2 (November 1957).

In, Chong Sik. 印貞植 *Chōsen no nōgyō kikō* 朝鮮の農業機構 (Agricultural structure of Korea). Tokyo, 1940.

Institute of Pacific Relations. *Industrial Japan: Aspects of Economic Changes as Viewed by Japanese Writers.* Compiled and translated by the Research Staff of the Secretariat. New York, 1941.

Itani, Zen'ichi 猪谷善一 *Chōsen keizai-shi* 朝鮮経済史 (Economic history of Korea). Tokyo, 1928.

Iwakata, Isō 岩片磯雄 *Chōsen-mai seisan-hi ni kansuru chōsa* 朝鮮米生産費に関する調査 (Survey of the cost of rice production in Korea). Tokyo, Japanese Society for the Promotion of Scientific Research, 1936.

Johnston, Bruce H. "Agricultural Productivity and Economic Development in Japan," *Journal of Political Economy* VIII.6 (December 1951).

———. *Japanese Food Management in World War II.* Stanford, Stanford University Press, 1953.

Kajikawa, Hansaburō 梶川半三郎 *Jitsugyō no Chōsen* 実業の朝鮮 (Industrial Korea). Seoul, Society for the Study of Korea, 1911.

Kada, Naoji 賀田直治 *Chōsen kōgyō kihon chōsa gaiyō* 朝鮮工業基本調査概要 (Review of basic studies on Korean industry). Seoul, 1934.

Kamada, Sawaichirō 鎌田澤一郎 *Tekunokurashī to Chōsen shigen no hiyaku* テクノクラシーと朝鮮資源の飛躍 (Technocracy and the development of Korean resources). Seoul, 1933.

Kawada, Shirō. 河田嗣郎 "Nihon to Chōsen ni okeru

kosakunin-seido" 日本と朝鮮に於ける小作人制度 (Tenant system in Japan and Korea), *Kyoto Teikoku Daigaku Keizai ronshū* Vol. I (July 1926).

———. "Chōsen ni okeru nōgyō shin'yō-gashi" 朝鮮に於ける農業信用貸 (Agricultural credit in Korea), *Kyoto Teikoku Daigaku Keizai ronshū,* Vol. II (July 1927).

Kawai, Akitake 川合彰武 *Chōsen kōgyō no gendankai* 朝鮮工業の現段階 (The current stage of Korean manufacturing). Seoul, 1943.

Keijō Nippō sha 京城日報社 (Seoul Daily News Co.). *Chōsen Nenkan* 朝鮮年鑑 (Yearbook of Korea). 1930–1940.

Keijō Shōkō Kaigi-sho 京城商工会議所 (Seoul Chamber of Commerce). *Chōsen no kōsan to kōjō* 朝鮮の工産と工場 (Industrial products and factories in Korea). Seoul, 1927.

Kindleberger, Charles P. *Foreign Trade and the National Economy.* New Haven, Yale University Press, 1962.

———. *Economic Development.* New York, McGraw-Hill Book Company, 1965.

Ko, Sung-je 高承濟 *Kensei Hanguk sanop sa* 近世韓國產業史 (Modern history of Korean industries). Seoul, 1950.

———. *Hanguk kyungje ron* 韓國經濟論 (The Korean economy). Seoul, 1956.

Kobayakawa, Kurō, ed. 小早川九郎 *Chōsen nōgyō hattatsu-shi* 朝鮮農業発達史 (History of agricultural development in Korea). Vol. I: *Development;* Vol. II: *Policy.* Seoul, Agricultural Association of Korea, 1944.

Kuznets, Simon. *Commodity Flow and Capital Formation.* 2 vols. New York, National Bureau of Economic Research, 1938.

———. *National Income and Its Composition.* 2 vols. New York, National Bureau of Economic Research, 1941.

———. "Long-Term Changes in the National Income of the United

States of America Since 1870," in *Income and Wealth of the United States, Trends and Structure,* ed. Simon Kuznets. Income and Wealth Series, Vol. II. Baltimore, The Johns Hopkins University Press, 1952.

———. *Six Lectures on Economic Growth.* Glencoe, Illinois, Free Press, 1960.

———. *Capital in the American Economy, Its Formation and Financing.* Studies in Capital Formation and Financing, National Bureau of Economic Research. Princeton, Princeton University Press, 1961.

———. *Postwar Economic Growth: Four Lectures.* Cambridge, The Belknap Press of Harvard University, 1964.

———. *Economic Growth and Structure, Selected Essays.* New York, W. W. Norton & Co., Inc., 1965.

———. "Quantitative Aspects of the Economic Growth of Nations," *Economic Development and Cultural Change* V.1 (October 1956); V.4 (July 1957); VI.4, Part 2 (July 1958); VII.3, Part 2 (April 1959), VIII.4, Part 2 (July 1960); I.3, Part 2 (April 1961); IX.4, Part 2 (July 1961); X.2, Part 2 (January 1962); XI.2, Part 2 (January 1963).

Kuznets, Simon, W. E. More, and J. J. Spengler, eds. *Economic Growth: Brazil, India, Japan.* Durham, Duke University Press, 1955.

Lee, Chang-nyol 李昌烈 *Hanguk kyūngje kucho wa sunwhan* 韓國經濟構造와循環 (Structure and flow of the Korean economy). Seoul, 1958.

Lee, Hoon K. *Land Utilization and Rural Economy in Korea.* Hong Kong, Kelly and Walsh, Ltd., 1936.

Lee, Ki Baek 李基伯 *Hanguk sa sin lon* 韓國史新論 (New interpretations of Korean history). Seoul, 1969.

Lee, Man-gi. 李滿基 *Hanguk kyūngje ron* 韓國經濟論

(The Korean economy). Seoul, 1963.

Lewis W. Arthur, *The Theory of Economic Growth.* Homewood, Richard D. Irwin, Inc., 1955.

Lockwood, William W. *The Economic Development of Japan: Growth and Structural Change, 1868–1938.* Princeton, Princeton University Press, 1954.

———. *The State and Economic Enterprise in Japan.* Studies in the Modernization of Japan. Princeton, Princeton University Press, 1965.

Mason, Edward S. *Economic Planning in Underdeveloped Areas: Government and Business.* The Millar Lectures, No. 2. New York, Fordham University Press, 1958.

McCune, George H. *Korea Today.* Cambridge, Harvard University Press, 1950.

McCune, Shannon. *Korea's Heritage: A Regional and Social Geography.* Tokyo, 1956.

Miyake, Shikanosuke 三宅鹿之助 "Chōsen to naichi to no keizai-teki kankei" 朝鮮と内地との経済的関係 (Economic relations between Korea and Japan), *Keijō Teikoku Daigaku Hōbun Gakkai: dai-ichi ronshū* 京城帝国大学法文学会第一論集 (Review of the Society of Law and Letters, Keijō Imperial University, First Series). Seoul, 1928.

Namchōsen Kwado Chonygbū 南朝鮮過渡政府 (The Provisional Government of South Korea). *Chōsun t'onggye nyungam* 朝鮮統計年鑑 1943 (Statistical yearbook of Korea, 1943). Seoul, 1948.

Nathan, Robert and Associates. *An Economic Programme for Korean Reconstruction.* Prepared for the United Nations Korean Reconstruction Agency. New York, 1954.

National Bureau of Economic Research. *The Comparative Study of Economic Growth and Structure: Suggestions on Research Objectives and Organizations.* New York, 1959.

Nelson, M. Frederick. *Korea and the Old Order in Eastern Asia.* Baton Rouge, Louisiana State University Press, 1946.

Nihon Tōkei Kenkyūjō 日本統計研究所 (Statistical Research Institute of Japan). *Nihon keizai tōkeishū* 日本経済統計集 (Japanese economic statistics). Tokyo, 1958.

Oda, Tadao 小田忠夫 "Keijō shoki ni okeru Chōsen Sōtokufu keizai no hattatsu" 京城初期に於ける朝鮮総督府経済の発達 (The development of public finance in the early years of the administration of the Government-General), *Chōsen keizai no kenkyū* 朝鮮経済の研究 (Studies in the Korean economy), Keijō Teikoku Daigaku Hōgakkai ronshū. Tokyo, 1938.

Ohkawa, Kazushi and Henry Rosovsky. "The Role of Agriculture in Modern Japanese Economic Development," *Economic Development and Cultural Change* IX.1, Part 2 (October 1960).

———. "A Century of Japanese Growth," in *The State and Economic Enterprise in Japan,* ed. W. W. Lockwood. Princeton, Princeton University Press, 1965.

Ohkawa, Kazushi in association with M. Shinohara, M. Umemura, M. Ito, and T. Noda, *The Growth Rate of the Japanese Economy Since 1878.* Tokyo, 1957.

Ōuchi, Takeji 大内武次. "Chōsen ni okeru beikoku seisan," 朝鮮に於ける米穀生産 (The production of rice in Korea), *Chōsen keizai no kenkyū* 朝鮮経済の研究 (Studies in the Korean economy). Tokyo, 1936.

Paek Yong-hun 白永勳. *Hanguk kyūngje wa gongup e kwanhan yongu* 韓國經濟와工業化關한研究 (Studies in the Korean economy and industrial development). Seoul, 1962.

Princeton University School of Public Affairs and the Population Association of America. *Population Index* X.4 (October 1944); XVI.4 (October 1950).

Ranis, Gustav. "Factor Proportions in Japanese Economic Development," *American Economic Review* XLVII.5 (September 1957).

———. "The Capital Output Ratio in Japanese Economic Development," *Review of Economic Studies* XXVI.5 (October 1958).

———"The Financing of Japanese Economic Development," *Economic History Review* XI.3 (April 1959).

Republic of Korea Ministry of Reconstruction. *Economic Survey.* 1958, 1959.

Republic of Korea Economic Planning Board. *Summary of the First Five-Year Economic Plan.* 1962–1966.

Rosovsky, Henry. "The Statistical Measurement of Japanese Economic Growth," *Economic Development and Cultural Change* VII.1 (October 1958).

———*Capital Formation in Japan, 1868–1940.* Glencoe, Illinois, Free Press, 1961.

———, ed. *Industrialization in Two Systems,* Essays in Honor of Alexander Gerschenkron. New York, John Wiley & Sons, Inc., 1966.

Rostow, W. W. *The Stages of Economic Growth.* Cambridge, The Cambridge University Press, 1961.

———, ed. *The Economics of Take-Off into Sustained Growth.* Proceedings of a conference held by the International Economic Association. New York, St. Martin's Press, Inc., 1963.

Schultz, Theodore W. *Transforming Traditional Agriculture.* New Haven, Yale University Press, 1964.

———"Investment in Human Capital," *American Economic Review* LI.1 (March 1961).

Schumpeter, E. B., ed. *The Industrialization of Japan and Manchuria, 1930–1940.* New York, The Macmillan Company, 1940.

Seoul Hanguk-Nyunkam Pyeunchanhoe 서울 韓國年鑑編纂會 (Seoul Korean Yearbook Committee). *Hanguk-nyungam* 韓國年鑑 (Yearbook of Korea). 1955–1962.

Shikata, Hiroshi 四方博. "Chōsen ni okeru kindai shihonshugi no seiritsu katei," 朝鮮に於ける近代資本主義の成立過程 (The process of formation of modern capitalism in Korea), *Chōsen shakai keizai-shi kenkyū* 朝鮮社会経済史研究 (Studies in Korean social and economic history). *Keijō Teikoku Daigaku Hōgakkai ronshū,* (Keijo Imperial University Law Society Review), Tokyo, 1933.

———"Richō jinkō ni kansuru mibun kaikyū-betsuteki kansatsu," 李朝人口に関する身分階級別的観察 (A study of social classes as related to population during the Yi dynasty), *Chōsen keizai no kenkyū* 朝鮮経済の研究 (Studies in Korean economy). *Keijō Teikoku Daigaku Hōgakkai ronshū,* Tokyo, 1938.

———"Shijō o tsūjite mitaru Chōsen no keizai" 市場を通じて見たる朝鮮の経済 (The Korean economy from the standpoint of the market), *Chōsen keizai no kenkyū. Keijō Teikoku Daigaku Hōbun Gakkai: dai-ichi-bu ronshū* 朝鮮経済の研究·京城帝国大学法文学会·第一部論集 Tokyo, 1939.

Smith, Thomas C. *The Agrarian Origins of Modern Japan.* Stanford, Stanford University Press, 1959.

Sung Chang-whan 成昌煥. *Hanguk kyūngje ron* 韓國經濟論 (The Korean economy). Seoul, 1956.

Suzuki, Seibun 鈴木成文. *Chōsen keizai no gendankai* 朝鮮經済の現段階 (The current stage of the Korean economy). Seoul, 1930.

Suzuki, Takeo 鈴木武雄. "Richō makki ni okeru Chōsen no

keizai" 李朝末期に於ける朝鮮の経済 (Public finance in the latter part of the Yi dynasty), *Chōsen keizai no kenkyū* 朝鮮経済の研究 (Studies in Korean economy). *Keijō Teikoku Daigaku Hōbun Gakkai: dai-ichi-bu ronshū* 京城帝国大学法文学会. 第一部論集 (Keijo Imperial University Review of the Society of Law and Letters). Tokyo, 1929.

———*Chōsen kin'yū-ron jikkō* 朝鮮金融論十講 (Ten lectures on the Korean banking system). Seoul, 1940.

———*Chōsen keizai no shinkōsō* 朝鮮経済の新構想. (A new conception of the Korean economy). Tokyo, 1942.

———*Chōsen no keizai* 朝鮮の経済 (The Korean economy). Tokyo, 1942.

Taehan Minguk 大韓民國 (Republic of Korea). *Taehan Minguk Cheilhoe Chong-ingu chosa kyolgwa sokpo* 大韓民國第一回總人口調査結果速報 (Preliminary report of the first census of the Republic of Korea). Seoul, 1949: *Taehan Minguk kani chong-ingu chosa* 大韓民國簡易總人口調査 (Population census report of Korea). Seoul, 1955: *Inku chutaik kuksei chosa bogo* 人口住宅國勢調査 *1960 (1960 Population and Housing Census in Korea).* Vol. II - *20 percent Sample Tabulation Report.*

———*Hanguk kyūngje tonggye wolbo* 韓國經濟統計月報 (Korean economic statistics, monthly summation). 1953–1960.

———*Hanguk tonggye nyungam* 韓國統計年鑑 (Statistical yearbook of Korea).

Title varies: *Taehan Minguk tonggye nyungam* 大韓民國統計年鑑, 1952–1960; *Hanguk tonggye nyungam* 韓國統計年鑑, 1961–1963.

———*Hanguk tonggye wolbo* 韓國統計月報 *(Monthly Statistics of Korea).* 1953–1960.

Title varies: *Minguk tonggye wolbo* 民國統計月報, 1961–.

———*Nodongryok chosa* 労働力調査 (Labor force survey). 1958–1960. *T'onggye chungbo* 統計情報 (Statistical Reporter).

Taeuber, Irene B. "Population and Labor Force in Industrializing Japan, 1850–1950," in *Economic Growth: Brazil, India, Japan,* Simon Kuznets, W. E. More, and J. J. Spengler, eds. Durham, Duke University Press, 1955.

———and George W. Barclay. "Korea and the Koreans in the Northeast Asian Region," *Population Index* XVI.4 (October 1950).

Takahashi, Kamekichi 高橋亀吉 *Gendai Chōsen keizairon* 現代朝鮮経済論 (The contemporary Korean economy). Tokyo, 1955.

Thomas, Warren B. *Population and Peace in the Pacific.* Chicago, University of Chicago Press, 1946.

———*Population and Progress in the Far East.* Chicago, University of Chicago Press, 1959.

Tōhata, Seiichi, 東畑精一 and Ōkawa Kazushi 大川一司 Chōsen beikoku keizairon 朝鮮米穀経済論 (Treatise on the rice economy of Korea). Tokyo, 1939.

Tōyō Takushoku Kabushiki Kaisha 東洋拓殖株式会社 (Oriental Development Company). *Nijū-nen-shi* 二十年史 (Twenty-year history). Tokyo, 1928.

Tsumagari, Kuranozō 津曲蔵之丞 "Chōsen ni okeru kosaku mondai no hatten katei," 朝鮮に於ける小作問題の発展過程 (Development of the tenancy problem in Korea *Chōsen keizai no kenkyū* 朝鮮経済の研究 (Studies in Korean economy). *Keijō Teikoku Daigaku Hōbun Gakubu, Hōbun Gakkai dai'ichi-bu ronshū.* Tokyo, 1929.

United Nations, Department of Social Affairs, Population Division. *Population Bulletin No. 1.* December 1951.

———*Future Population Estimated by Sex and Age: Report IV, The Population of Asia and the Far East, 1950–1960.*

———*Demographic Aspects of Manpower: Report I, Sex and Age Pattern of Participation in Economic Activities.*

———*Population Studies No. 33, The Determinants and Consequences of Population Trend.* New York, 1953.

United Nations Economic Commission for Asia and the Far East. *Economics Survey of Asia and the Far East, 1948–1962.*

———*Economic Bulletin for Asia and the Far East, 1950–1962.*

United Nations Food and Agricultural Organization. *Rehabilitation and Development of Agriculture: Forestry and Fisheries in South Korea.* Report prepared for the United Nations Korean Reconstruction Agency. New York, Columbia University Press, 1954.

Universities-National Bureau Committee Economic Research. *Problems in the Study of Economic Growth.* New York, 1949.

Watanabe, T. "Economic Aspects of Dualism in the Industrial Development of Japan, *Economic Development and Cultural Change* (April 1965).

Won Yonk-sok 元容奭 *Junran jung nongup kyungje* 戰亂中農業經濟 (Agricultural economy during the Korean War). Seoul, 1953.

Yamada, Fumio 山田文雄. "Chōsenjin rōdō-sha mondai" 朝鮮人労働者問題 (Problems of Korean laborers), *Chōsen keizai no kenkyū* 朝鮮経済の研究 (Studies in Korean economy). *Keijō Teikoku Daigaku Hōbun Gakkai: dai-ichi-bu ronshū.* Tokyo, 1929.

Yamada, Yūzō 山田雄三. *Nihon kokumin shotoku suikei shiryō* 日本国民所得推計資料 (A comprehensive survey of national income data in Japan). Tokyo, 1937.

Yamaguchi, Toyomasa 山口豊正 *Chōsen no kenkyū* 朝鮮の研究 (Study of Korea). Tokyo, 1911.

Zenkoku Keizai Chōsa Kikan Rengōkai, Chōsen Shibu 全国経済調査機関联合会,朝鮮支部 (National Federation of Economic Research Organizations, Korean Branch). *Chōsen keizai nempō* 朝鮮経済年報 (Economic annual of Korea). 1939, 1940, 1941–1942.

Zenshō, Eisuke 善生永助. *Chōsen no ichiba* 朝鮮の市場 (Korean markets). Seoul, 1927.

———Chōsen no jinkō genshō 朝鮮の人口現象 (Population trend in Korea). Seoul, 1927.

Other Bibliographies

Changso Mokrok 藏書目録 (Classified Catalogue of Books in the National Assembly Library of Korea). Seoul, 1961.

Eackus, Robert L. *Russian Supplement to the Korean Studies Guide.* Berkeley, The University of California Press, 1958.

R. H. Hazard Jr., James Hoyt, H. T. Kim, and W. W. Smith Jr., for the Institute of East Asiatic Studies, University of California. *Korean Studies Guide.* Berkeley and Los Angeles, The University of California Press, 1954.

Korea University, Asiatic Research Center. *Bibliography of Korean Studies:* A Bibliographical Guide to Korean Publications on Korean Studies Appearing from 1945 to 1958. Seoul, 1961. A similar guide for the years 1959 to 1962. Seoul, 1965.

Korean National Library. *Taehan Minkuk Chulpanmul Chong Mokrok* 大韓民國出版物總目録 (Korean National Publication Bibliography). Covering 1945–1962, Seoul, 1964. Covering 1963–1964, Seoul, 1965.

U.S. Library of Congress. *Korea: An Annotated Bibliography of Publications in the Russian Languages.* Washington, 1950.

U.S. Library of Congress. *Korea: An Annotated Bibliography of Publications in Western Languages.* Washington, 1950.

U.S. Library of Congress. *Korea: An Annotated Bibliography of*

Publications in Far Eastern Languages. Washington, 1950.
Yonsei University, Industrial Economics Research Institute. *Sanop Munhun Mokrok, 1945–1960* 產業文獻目錄 (Catalogue of Publications on Korean Industries, 1945–1960). Seoul, 1961.

INDEX

HARVARD EAST ASIAN MONOGRAPHS

1. Liang Fang-chung, *The Single-Whip Method of Taxation in China*
2. Harold C. Hinton, *The Grain Tribute System of China, 1845-1911*
3. Ellsworth C. Carlson, *The Kaiping Mines, 1877-1912*
4. Chao Kuo-chün, *Agrarian Policies of Mainland China: A Documentary Study, 1949-1956*
5. Edgar Snow, *Random Notes on Red China, 1936-1945*
6. Edwin George Beal, Jr., *The Origin of Likin, 1835-1864*
7. Chao Kuo-chün, *Economic Planning and Organization in Mainland China: A Documentary Study, 1949-1957*
8. John K. Fairbank, *Ch'ing Documents: An Introductory Syllabus*
9. Helen Yin and Yi-chang Yin, *Economic Statistics of Mainland China, 1949-1957*
10. Wolfgang Franke, *The Reform and Abolition of the Traditional Chinese Examination System*
11. Albert Feuerwerker and S. Cheng, *Chinese Communist Studies of Modern Chinese History*
12. C. John Stanley, *Late Ch'ing Finance: Hu Kuang-yung as an Innovator*
13. S. M. Meng, *The Tsungli Yamen: Its Organization and Functions*
14. Ssu-yü Teng, *Historiography of the Taiping Rebellion*
15. Chun-Jo Liu, *Controversies in Modern Chinese Intellectual History: An Analytic Bibliography of Periodical Articles, Mainly of the May Fourth and Post-May Fourth Era*
16. Edward J. M. Rhoads, *The Chinese Red Army, 1927-1963: An Annotated Bibliography*

17. Andrew J. Nathan, *A History of the China International Famine Relief Commission*

18. Frank H. H. King (ed.) and Prescott Clarke, *A Research Guide to China-Coast Newspapers, 1822-1911*

19. Ellis Joffe, *Party and Army: Professionalism and Political Control in the Chinese Officer Corps, 1949-1964*

20. Toshio G. Tsukahira, *Feudal Control in Tokugawa Japan: The Sankin Kōtai System*

21. Kwang-Ching Liu, ed., *American Missionaries in China: Papers from Harvard Seminars*

22. George Moseley, *A Sino-Soviet Cultural Frontier: The Ili Kazakh Autonomous Chou*

23. Carl F. Nathan, *Plague Prevention and Politics in Manchuria, 1910-1931*

24. Adrian Arthur Bennett, *John Fryer: The Introduction of Western Science and Technology into Nineteenth-Century China*

25. Donald J. Friedman, *The Road from Isolation: The Campaign of the American Committee for Non-Participation in Japanese Aggression, 1938-1941*

26. Edward Le Fevour, *Western Enterprise in Late Ch'ing China: A Selective Survey of Jardine, Matheson and Company's Operations, 1842-1895*

27. Charles Neuhauser, *Third World Politics: China and the Afro-Asian People's Solidarity Organization, 1957-1967*

28. Kungtu C. Sun, assisted by Ralph W. Huenemann, *The Economic Development of Manchuria in the First Half of the Twentieth Century*

29. Shahid Javed Burki, *A Study of Chinese Communes, 1965*

30. John Carter Vincent, *The Extraterritorial System in China: Final Phase*

31. Madeleine Chi, *China Diplomacy, 1914-1918*

32. Clifton Jackson Phillips, *Protestant America and the Pagan World: The First Half Century of the American Board of Commissioners for Foreign Missions, 1810-1860*

33. James Pusey, *Wu Han: Attacking the Present through the Past*

34. Ying-wan Cheng, *Postal Communication in China and Its Modernization, 1860-1896*

35. Tuvia Blumenthal, *Saving in Postwar Japan*

36. Peter Frost, *The Bakumatsu Currency Crisis*

37. Stephen C. Lockwood, *Augustine Heard and Company, 1858-1862*

38. Robert R. Campbell, *James Duncan Campbell: A Memoir by His Son*

39. Jerome Alan Cohen, ed., *The Dynamics of China's Foreign Relations*

40. V. V. Vishnyakova-Akimova, *Two Years in Revolutionary China, 1925-1927*, tr. Steven I. Levine

41. Meron Medzini, *French Policy in Japan during the Closing Years of the Tokugawa Regime*

42. *The Cultural Revolution in the Provinces*

43. Sidney A. Forsythe, *An American Missionary Community in China, 1895-1905*

44. Benjamin I. Schwartz, ed., *Reflections on the May Fourth Movement: A Symposium*

45. Ching Young Choe, *The Rule of the Taewŏn'gun, 1864-1873: Restoration in Yi Korea*

46. W. P. J. Hall, *A Bibliographical Guide to Japanese Research on the Chinese Economy, 1958-1970*

47. Jack J. Gerson, *Horatio Nelson Lay and Sino-British Relations, 1854-1864*

48. Paul Richard Bohr, *Famine and the Missionary: Timothy Richard as Relief Administrator and Advocate of National Reform*

49. Endymion Wilkinson, *The History of Imperial China: A Research Guide*

50. Britten Dean, *China and Great Britain: The Diplomacy of Commercial Relations, 1860-1864*

51. Ellsworth C. Carlson, *The Foochow Missionaries, 1847-1880*

52. Yeh-chien Wang, *An Estimate of the Land-Tax Collection in China, 1753 and 1908*

53. Richard M. Pfeffer, *Understanding Business Contracts in China, 1949-1963*

54. Han-sheng Chuan and Richard Kraus, *Mid-Ch'ing Rice Markets and Trade, An Essay in Price History*

55. Ranbir Vohra, *Lao She and the Chinese Revolution*

56. Liang-lin Hsiao, *China's Foreign Trade Statistics, 1864-1949*

57. Lee-hsia Hsu Ting, *Government Control of the Press in Modern China, 1900-1949*

58. Edward W. Wagner, *The Literati Purges: Political Conflict in Early Yi Korea*

59. Joungwon A. Kim, *Divided Korea: The Politics of Development, 1945-1972*

60. Noriko Kamachi, John K. Fairbank, and Chūzō Ichiko, *Japanese Studies of Modern China Since 1953: A Bibliographical Guide to Historial and Social-Science Research on the Nineteenth and Twentieth Centuries, Supplementary Volume for 1953-1969*

61. Donald A. Gibbs and Yun-chen Li, *A Bibliography of Studies and Translations of Modern Chinese Literature, 1918-1942*

62. Robert H. Silin, *Leadership and Values: The Organization of Large-Scale Taiwanese Enterprises*

63. David Pong, *A Critical Guide to the Kwangtung Provincial Archives Deposited at the Public Record Office of London*

64. Fred W. Drake, *China Charts the World: Hsu Chi-yü and His Geography of 1848*

65. William A. Brown and Urgunge Onon, translators and annotators, *History of the Mongolian People's Republic*

66. Edward L. Farmer, *Early Ming Government: The Evolution of Dual Capitals*

67. Ralph C. Croizier, *Koxinga and Chinese Nationalism: History, Myth, and the Hero*

68. William J. Tyler, tr., *The Psychological World of Natsumi Sōseki*, by Doi Takeo

69. Eric Widmer, *The Russian Ecclesiastical Mission in Peking during the Eighteenth Century*

70. Charlton M. Lewis, *Prologue to the Chinese Revolution: The Transformation of Ideas and Institutions in Hunan Province, 1891-1907*

71. Preston Torbert, *The Ch'ing Imperial Household Department: A Study of its Organization and Principal Functions, 1662-1796*

72. Paul A. Cohen and John E. Schrecker, eds., *Reform in Nineteenth-Century China*

73. Jon Sigurdson, *Rural Industrialization in China*

74. Kang Chao, *The Development of Cotton Textile Production in China*

75. Valentin Rabe, *The Home Base of American China Missions, 1880-1920*

76. Sarasin Viraphol, *Tribute and Profit: Sino-Siamese Trade, 1652–1853*

77. Ch'i-ch'ing Hsiao, *The Military Establishment of the Yuan Dynasty*

78. Meishi Tsai, *Contemporary Chinese Novels and Short Stories, 1949–1972: An Annotated Bibliography*

79. Wellington K. K. Chan, *Merchants, Mandarins, and Modern Enterprise in Late Ch'ing China*

80. Endymion Wilkinson, *Landlord and Labor in Late Imperial China: Case Studies from Shandong by Jing Su and Luo Lun*

81. Barry Keenan, *The Dewey Experiment in China: Educational Reform and Political Power in the Early Republic*

82. George A. Hayden, *Crime and Punishment in Medieval Chinese Drama: Three Judge Pao Plays*